Mechanical Engineering Series

Series Editor

Vish Prasad, Department of Mechanical and Energy Engineering, University of North Texas, Denton, TX, USA

The Mechanical Engineering Series presents advanced level treatment of topics on the cutting edge of mechanical engineering. Designed for use by students, researchers and practicing engineers, the series presents modern developments in mechanical engineering and its innovative applications in applied mechanics, bioengineering, dynamic systems and control, energy, energy conversion and energy systems, fluid mechanics and fluid machinery, heat and mass transfer, manufacturing science and technology, mechanical design, mechanics of materials, micro- and nano-science technology, thermal physics, tribology, and vibration and acoustics. The series features graduate-level texts, professional books, and research monographs in key engineering science concentrations.

Sebastian Volz • Jose Ordonez-Miranda

Heat Transport Driven by Surface Electromagnetic Waves

Springer

Sebastian Volz
LIMMS CNRS – The University of Tokyo
Tokyo, Japan

Jose Ordonez-Miranda
LIMMS CNRS – The University of Tokyo
Tokyo, Japan

ISSN 0941-5122 ISSN 2192-063X (electronic)
Mechanical Engineering Series
ISBN 978-3-031-66023-8 ISBN 978-3-031-66024-5 (eBook)
https://doi.org/10.1007/978-3-031-66024-5

This Springer imprint is published by the registered company Springer Nature Switzerland AG
The registered company address is: Gewerbestrasse 11, 6330 Cham, Switzerland

To my mentors, Michel Lallemand, Jean-Bernard Saulnier, Gang Chen, and Jean-Jacques Greffet. (SV)

To the memory of my grandmother, Eudocia Miranda Hurtado. (JOM)

Foreword

I learned molecular dynamics from Sebastian Volz when he did a post-doc with me between 1997 and 1998 and surface phonon polaritons from Jean-Jacques Greffet in 2000 at a conference in Banff. Sebastian joined Centre National de la Recherche Scientifique (CNRS) after his postdoc and resided in Jean-Jacques group at Ecole Centrale Paris. Around that time, the metamaterials field emerged, and I started working on surface phonon polaritons, leveraging the negative dielectric constant of surface phonon polaritons in the infrared regime, which is more relevant to heat transfer. The major direction was near-field radiation heat transfer through nanoscale gaps mediated by surface phonon polaritons, first predicted by Jean-Jacques group. Our experiments, led by several brave students, Lu Hu, Arvind Narayanaswamy, and Shen Sheng led to some initial successes.

Another PhD student at the time, Dye-Zone (Zony) Chen, was looking for a PhD research topic. By then, it has been well-known that the phonon thermal conductivity reduces in thin films due to boundary scattering. I was looking for ways to reverse this trend. Knowing that the high density of states was behind the increased near-field radiation heat transfer between two bodies separated by a narrow gap, I was curious if the same effect could increase the in-plane thermal conductivity of thin films, and suggested Zony looking into this direction, with the support of Arvind Narayanaswamy. Zony's study, however, found that the increased density of states does not contribute much to the effective thermal conductivity of thin films. Rather, it was the long propagation length of the asymmetric mode of surface phonon polaritons that can potentially lead to an appreciable effective thermal conductivity. In 2005, we published our first paper on this topic in Physical Review B. After Zony wrapped up his PhD thesis, we moved to other topics. Little did I expect that subsequent studies would make this a very interesting topic.

Sebastian and his collaborators, including the junior author of this monograph Jose Ordonez-Miranda, who started as a post-doctoral fellow with Sebastian, carried out extensive experiments and modeling to explore the impact of surface phonon polaritons on heat conduction. Their experiments demonstrated that the surface phonon polaritons can indeed contribute appreciably to the heat conduction along thin films, and this contribution becomes higher for thinner films. The surface phonon polariton contribution also increases with increasing temperature. Their extensive theoretical modeling on thin films and nanowires adds significantly to our understanding. The agreements between modeling results and experiments build confidence in the idea that surface phonon polaritons can make meaningful contributions to heat conduction at small scale.

Meanwhile, other groups also reported some interesting, but still puzzling observations: experiments have shown significantly enhanced thermal conductivity even for relatively short films or nanowires. These include the work of Deyu Li from Vanderbilt and Renkun Chen from UC San Diego, and others around the world. Different explanations for such anomalous increase of thermal conductivity in very thin films and nanowires have been given, including the hypothesis of nonequilibrium transport. It is increasing to note that the work of Xianfan Xu from Purdue showed that the hyperbolic dispersion in anisotropic materials can lead to phonon polariton contributions comparable to phonons.

There seems to be something we do not fully understand yet about surface phonon polaritons. The topic is interesting because phonon polaritons provide a bridge between heat conduction and thermal radiation. By nature, phonon polaritons are hybridization of phonons, which are most responsible for heat conduction, and photons, which are most responsible for radiation.

Sebastian and Jose's monograph is a timely one as it is a comprehensive summary on what we know. The monograph starts from the fundamentals of thermal radiation, electromagnetic waves, and phonon polaritons, and explains clearly the wave to the classical particle picture transition. It then discusses the surface phonon polaritons in thin films and wire configurations, and thermal transport in the diffusive and the ballistic limits and in between the two limits. I found that the manograph is succinct and easy to follow, and I believe that it will be an excellent read for researchers who are interested in this topic. With continued exploration, new understandings and applications could emerge.

Massachusetts
May 2024

Gang Chen

Preface

This book represents the culmination of a decade-long research into the field of low-dimensional heat transport driven by surface polariton waves. Born from the seeds sown by Professor Gang Chen's seminal paper at MIT, this scientific pursuit has unraveled the less expected impact of polariton heat in polar, metal, and doped dielectric films. We compile here the first comprehensive exposition of this field at the triple point among electromagnetism, transport physics, and heat transfer. Dedicated to graduate students and researchers seeking to investigate the node of these three domains, this book intends to serve as a reference to understand basic polariton thermal transport and its microscopic origin.

Promises of polariton thermal radiation range across various domains of thermal management, sensing, and computing. Considering heat rectification, guiding, and tunneling, polariton waves offer unprecedented opportunities for controlling the flow of thermal energy. The wave features of polaritons allow for the manipulation, steering, and focusing of their heat fluxes. Moreover, their capacity to be integrated into circuits yields the possible realization of novel thermal rectifiers and tunable thermal diodes in the landscape of thermotronics. In low-dimensional materials, polariton thermal radiation is expected to exert significant effects in two-dimensional and one-dimensional systems, where surface confinement plays a pivotal role. Notably, the far-field emission of polariton radiation enables its utilization in long-range thermal transport, which is especially promising for remote sensing and energy harvesting. Furthermore, the inherent controllability of polariton thermal radiation makes it a versatile tool for manipulating heat at the nanoscale, paving the way for innovative solutions in nanoscale heat transfer. In essence, polariton thermal radiation represents a paradigm shift in our ability to harness and manipulate heat, offering new avenues for advancing thermal science and technology.

Spanning a trajectory from foundational principles to advanced technical insights, the book is structured to facilitate the reader's progression. Chapters 1 and 2 lay the groundwork with an exposition of basic concepts, while the subsequent chapters introduce deeper insights into the Maxwell's equations and dielectric functions (Chap. 3), dispersion relations (Chap. 4), and polariton thermal conductances (Chaps. 5 and 6) of nanofilms and nanowires. Each chapter includes self-consistent notations and symbols that might differ between chapters. The references cited herein reflect a balanced synthesis of both foundational papers and books on one side and topic-specific papers on the other side.

We are deeply indebted to the numerous collaborators who have supported and nourished this scientific journey. In alphabetic order, they are Thomas Antoni, Roman Anufriev, Sergei Gluchko, Byunggi Kim, Yuriy Kosevich, Laurent Jalabert, Masahiro Nomura, Laurent Tranchant, Saeko Tachikawa, Shiyun Xiong, and Yunhui Wu. Special gratitude is reserved for Professor Gang Chen and Professor Jean-Jacques Greffet, whose visionary insights and mentorship have encouraged us to contribute to the development of this research field. As we write these words, we are heartened by the growing interest in polariton heat transport, as evidenced by recent and multiple contributions especially from the USA and Asia. It is our fervent hope that this volume will serve as a guide towards new frontiers in the captivating field of low-dimensional radiation.

Tokyo, Japan
March 2024

Sebastian Volz
Jose Ordonez-Miranda

Acknowledgments

We acknowledge the support of the Core Research for Evolutional Science and Technology (CREST) program of the Japan Science and Technology Agency (Grant No. JPMJCR19I1).

Contents

Chapter 1
Introduction to Thermal Radiation and Surface Polaritons

1.1 Introduction

A historical backdrop sets the stage for our investigation, dating back to the seminal proposal of surface polaritons in 1956 by Ugo Fano, as documented in his publication [1]. Subsequent to this proposition, extensive scholarly endeavors have documented the intricacies of surface polaritons in various compendiums, books, and scholarly articles [2–5]. Despite the wealth of knowledge accumulated over the years, our work endeavors to chart a distinctive course. We aspire to scrutinize these modes through the lens of an innovative and original perspective—that of heat transfer, more specifically, the realm of thermal radiation.

Exploration of surface polaritons primarily stems from the optics and photonics communities, where researchers delve into the propagation of a single monochromatic wave governed by Maxwell's equations. This investigation incorporates antenna, wave guides, and laser technologies across varying scales. Notably, the scrutinized electromagnetic spectrum often excludes the Infrared (IR) range due to experimental complexities. Conversely, the community engaged in radiative heat transfer has traditionally operated within the frame of photon wave packet ray tracing, a method that suppresses information regarding the wave's phase. Nevertheless, this community has mastered the fine characterization and measurement of thermal radiation spectra, typically utilizing Fourier Transform Infrared Spectrometry. Furthermore, they have developed and established a strong knowledge to accurately predict radiative heat fluxes in transparent or semitransparent media using the Radiative Transfer Equation. This book endeavors to bridge these different perspectives and to foster a deeper understanding of a radiative heat transfer mechanism relying on resonant surface electromagnetic waves, commonly known as surface polaritons.

In adopting this novel approach, we aim to shed new light on the characteristics of surface polaritons within the context of thermal radiation. This path is clearly

S. Volz, J. Ordonez-Miranda, *Heat Transport Driven by Surface Electromagnetic Waves*, Mechanical Engineering Series, https://doi.org/10.1007/978-3-031-66024-5_1

unconventional as radiative heat transfer introduces a new dimension to the study of surface polaritons. A certain view of this combination has been uncovered and fostered for several decades [6] in the physics community addressing issues related to Casimir forces and near-field thermal radiation in small vacuum gaps between two plane surfaces [7].

Especially, since the 2000s, the investigation of this near-field heat flux has been pursued in the mechanical engineering community. In polar materials, the amplitude of this flux was found to exceed the conventional Planck's predictions by several orders of magnitude and to carry a very narrow spectrum of photons [8–11], opening new perspectives to heat management and photovoltaics [12, 13]. Referred to as a Super-Planckian contribution, this phenomenon found explanation through the analogy of tunneling, where modes of interacting bodies exchange quanta without the involvement of an intermediate medium possessing a lower number of modes. This surprisingly high enhancement is generated by the coupling between the evanescent tails of surface waves travelling along each surface. Those electromagnetic waves are especially resonant in the cases of polar materials and metals, carrying phonon–polaritons and plasmon–polaritons on their surfaces.

In 2002, a groundbreaking study [14] further demonstrated the coherent far-field radiation of those polaritons from surfaces designed with adapted gratings. In this case, the emitted flux remained within the limit of Planck's Law as the emitted electromagnetic waves have to comply with the mode structure in vacuum.

Later, another contribution of polariton radiation was discovered in the form of its in-plane heat transfer [15]. In polar films thinner than 100nm and at 500K, this radiative component would even overtake the vibrational heat conduction. This outcome deceives the common belief that radiative heat flux always remains smaller than the conductive one, at least by one order of magnitude. This theoretical work has been inspirational for more than a decade in the research path that the authors have been taking. This work is indeed at the basis of this new field tackling "two-dimensional" heat radiation along surfaces, interfaces, and films [16, 17] that we expose in detail in this book. This radiative contribution but in the one-dimensional transport was also predicted [18] and then experimentally proven [19]. Finally, it is remarkable that this contribution was also observed in bulk hyperbolic materials [20, 21].

1.1.1 General Comments on Photons, Phonons, and Electrons

These introductive comments serve as a basis for understanding the further developments on polaritons which conjugate photons with photons and photons with the electron gas. Well-established textbooks provide more in-depth coverage [22].

In quantum physics, particles or quasiparticles, such as photons, phonons, and electrons, emerge as fundamental quanta of energy. This quantum frame involves a profound transformation where classical electromagnetic and elastic fields are redefined in terms of quantum operators. These operators, in turn, define a

Hamiltonian that provides the solutions of the Schrödinger equation for each mode. Within this quantum framework, each mode becomes carrier of energy quanta, with the integer number of these particles ultimately dictating the amplitude and energy of the associated wave.

Photons are the source of thermal radiation. They define the quantum of electromagnetic energy, so they have only two transverse polarizations and are involved in interactions, especially with electrons. Their dispersion curve $\omega = ck/\sqrt{\epsilon}$, where ω, k, c, and ϵ stand, respectively, for the angular frequency, wave vector, light velocity, and dielectric function, passes through the origin and has no upper frequency limit. This dispersion reduces to that of vacuum or air when $\epsilon = 1$ to define the line or the light cone if the three dimensions of reciprocal space are taken into account. When the medium through which the light passes contains charges, the dielectric function is no longer unitary, and the dispersion deviates from that of the line of light. Nevertheless, the group velocity of light cannot exceed that in a vacuum.

Phonons are the quanta of vibrational waves in solids and, as such, are referred to as quasiparticles, since their nature is wave-like rather than corpuscular. Their dispersion is correctly obtained from the equations of elasticity in simple solids, but it is necessary to revert to the atomic equations of motion if the number of atoms per lattice cell increases. It is indeed possible to show that the number of branches is equal to three times the number of atoms per cell, and not all additional branches can be captured by the elasticity equations. Each branch is associated with a polarization, either longitudinal or transverse.

Phonon dispersion curves are represented in the first Brillouin zone, which is the smallest period of reciprocal space in which all modes are represented. Three branches pass through the origin and correspond to the acoustic modes, one of which is longitudinal and the other two transverse. The other branches are optical and do not pass through the origin. In simple solids, the dispersion relations of the acoustics mode are typically sinus functions where the frequency is proportional to the square root of the force constant. Phonons have a maximum frequency due to the finite-valued wave vectors defining the first Brillouin zone. This maximum frequency is generally given by the Debye frequency obtained from the group velocity of low-frequency longitudinal acoustic modes. This reference remains rather coarse, particularly when the dispersion of the modes is strong. The maximum frequency can be read directly from the complete dispersion curves and corresponds to that of the highest frequency optical mode. When the material is polar, certain optical modes, when they set charges of opposite sign in relative motion in the lattice cell, are said to be "active in the infrared," as they disturb the electromagnetic field. These modes are responsible for the formation of phonon–polaritons.

Electrons are charges, and their internal properties cannot be described by Maxwell's equations where charge is an input parameter. However, the potential for Coulombic interaction between charges, or between charge and ionic lattice, makes it possible to define a Hamiltonian and then to resolve the band structure of electrons in a solid. This dispersion relation is also defined in the first Brillouin zone as the symmetry of the crystal applies to electron modes. Electrons can now be described as waves, whereas their corpuscular nature was first demonstrated experimentally by the photoelectric effect. In a metal, electrons whose energy is higher than the Fermi energy and which belong to the conduction band are free: They circulate electric current and heat. Their dispersion relation is parabolic, their frequency being proportional to the wave vector squared. The collective and relative motion of the free electron gas toward the ion lattice generates a dipole at the origin of the plasmon–polariton modes.

1.1.2 Polaritons

The term "polariton" conveys the presence of an elementary particle emerging from the dynamics of a dipole. This dipole becomes manifest within specific solid-state materials. The movements of ions within polar materials lead to the emergence of phonon–polaritons, while the collective motions of electrons in metals result in the creation of a gas of negative charges, juxtaposed with the lattice ions. This collective behavior gives birth to entities known as plasmons and plasmon–polaritons, which play pivotal roles in the optical and thermal properties of materials.

To illustrate these complex phenomena, Figs. 1.1 and 1.2 provide a visual representation of the dipolar movements associated with polaritons. In Fig. 1.1, the dynamic movements of ions, which contribute to the formation of phonon–polaritons, are illustrated. These movements, associated with optical lattice vibrations, create an interplay between charged particles, leading to the emergence of local dipoles perturbating the electromagnetic field forming, in turn, these hybrid excitations of phonon–polaritons. Figure 1.2, on the other hand, illustrates the collective movements of electrons, showcasing the formation of a gas of negative

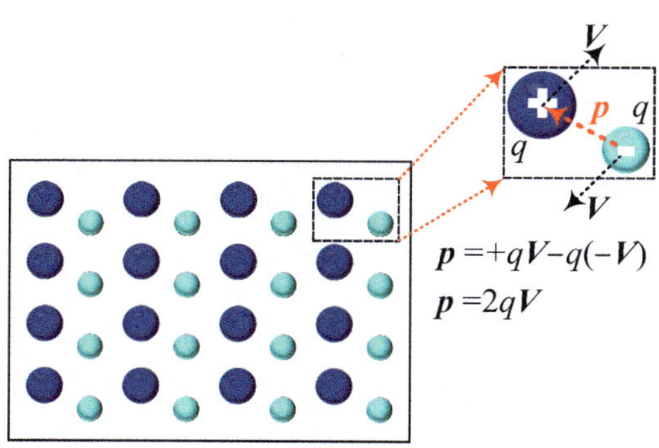

Fig. 1.1 Crystal lattice of a polar material made of two atoms per cell. This example is a toy model of a two-dimensional square lattice. Inset: schematic of a lattice cell including two ions of opposite signs. In a motion generated by optical modes, both atoms move in opposite directions, thus activating a dipolar excitation, which couples to the electromagnetic field. This hybrid excitation is called "phonon–polariton." The p vector refers here to the polarization and V to the atom velocity, while q denotes the charge

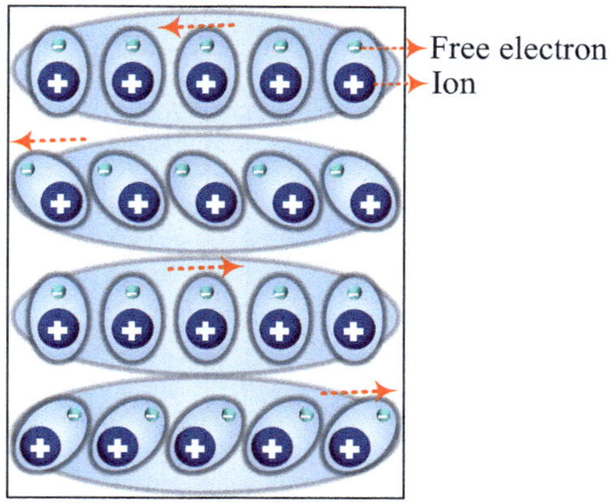

Fig. 1.2 The oscillation of the electron gas around the ion crystal due to the restoring force between electrons and the ion lattice produces a dipole coupled to the electromagnetic field. The resulting perturbation is called "plasmon–polariton," which mirrors the only longitudinal mode of the charge motions. The succession of layers corresponds to a sequence in time running from the upper layer to the bottom one

charges and the subsequent generation of plasmon–polaritons arising also from the coupling between the electromagnetic field and the ions–electron gas dipole.

Within the confines of this chapter, we will finally introduce the concept of surface polaritons. Our focus extends into the dynamic interplay of photons, phonons, and electrons shaping the fundamental nature of these phenomena. It is important to also underscore the existence of exciton polaritons, although their analysis will remain beyond the purview of our discussion.

1.2 Generalities on Thermal Radiation

1.2.1 Blackbody Radiation

The phenomenon of Blackbody thermal radiation [23] constitutes a cornerstone in our understanding of radiative heat transfer. Unveiling the fundamental laws that govern radiation indeed necessitates a foundational exploration starting with the definition of an idealized entity—the Blackbody. This perfect body is characterized by a maximal absorption, i.e., incident radiation encounters absolute absorption and zero reflection. Figure 1.3 provides a representation of a Blackbody through schematic illustrations and a depiction of its spectrum. The distinctive feature of a Blackbody radiation lies in its exclusive reliance on temperature to determine the nature of its emitted radiation.

The spectral flux density also called radiative intensity $I_0(\lambda)$ emitted by a Blackbody surface in a given direction and per surface unit is provided by Planck's Law:

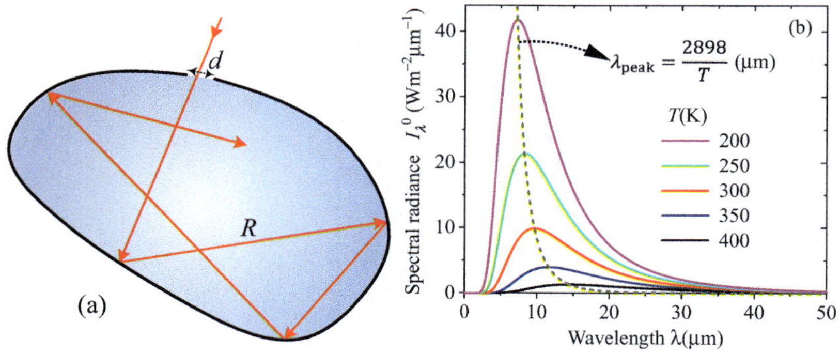

Fig. 1.3 (**a**) Device where a Blackbody surface is generated by opening a "pinhole" in a large cavity. In this configuration, any incoming ray is trapped, and the aperture can thus be considered as fully absorbing. The second Kirchhoff's Law then imposes the emissivity to be also at its maximum of unity. The cavity walls should be isothermal and set to the desired Blackbody temperature. (**b**) Spectral flux density also called radiative intensity emitted by a Blackbody surface as a function of wavelength according to Planck's Law. This latter directly provides the wavelength of maximum emission as reversely proportional to temperature as predicted by Wien's Law

$$I_\lambda^0 = \frac{2}{\lambda^5}\frac{hc^2}{\exp\left(\frac{hc}{\lambda k_B T}\right) - 1}, \tag{1.1}$$

where h, k_B, and T are, respectively, Planck's constant, the Boltzmann constant, and the Blackbody temperature in Kelvin. In Planck's function reported in Fig. 1.3b, the wavelength of maximum emission is established by Wien's Law as $\lambda_{peak} = 2897/T$ μm. It is remarkable that the Blackbody emission does not depend on direction, which makes of it a "Lambertian" body. It is also crucial to note that this law requires all dimensions of the system to be larger than the wavelengths involved, as stated by Planck himself in his Theory of Heat Radiation [24].

Consider, for instance, the Sun, a body with a temperature hovering around 6000 Kelvins. The characterization of the Sun as a Blackbody arises from the observation that any radiation incident upon its surface is predominantly absorbed, without reflection or transmission. The Sun emission thus occurs predominantly at a wavelength of approximately 500 nm, strategically positioned at the center of the visible spectral range. The Earth surface, at a temperature of about 300K, is emitting at its maximum at a wavelength of 10 μm. This difference in maxima is at the origin of the global climate warming, as greenhouse gases reduce the atmospheric transmission window around 10 μm.

As the Blackbody is Lambertian, the power density emanating from the Sun exhibits no dependence on direction. However, the incident power experienced by a receiver is intricately linked to its orientation and relative position concerning the solar source. This power can be found proportional to the sinus of the angle between the ray and the surface. The practical implications of those conditions resonate with our daily experiences on Earth. The tangible warmth felt during daylight hours and the subsequent cooling as the evening approaches are direct consequences of the changing angle of incidence of solar rays.

By establishing a robust relation between power, wavelength, temperature, and orientation, the Blackbody sets the reference for defining the emission of all other nonideal surfaces and bodies via the emissivity coefficient, being nothing else than the ratio between nonideal and blackbody fluxes. Furthermore, the second Kirchhoff's Law states, on the basis of a flux balance at the interface, that this spectral emissivity ϵ_λ also equals the spectral absorption α_λ:

$$\alpha_\lambda = \epsilon_\lambda = \frac{I_\lambda}{I_\lambda^0}. \tag{1.2}$$

1.2.2 Radiative Transfer Equation

A comprehensive approach of thermal radiation is provided by the Radiative Transfer Equation. This governing equation is based on the particular physics of rays, invoking linear trajectories of wave packets deriving from the combination of electromagnetic waves.

Thermal radiation finds apt description through the lens of photon wave packets–particles that, akin to billiard balls, traverse rectilinear paths without interacting with others. This behavior forms the very foundation of the Radiative Transfer Equation that serves as the balance equation for flux density in a specified direction:

$$\nabla I_\lambda = \epsilon(\lambda)\left[I_\lambda^0 - I_\lambda\right] + \sigma_\lambda\left[\langle I_\lambda\rangle_\Omega - I_\lambda\right], \tag{1.3}$$

where σ_λ refers to the scattering coefficient and the scattering is here assumed isotropic. The variation of the intensity in the volume element expressed on the left-hand side of the equation is balanced by the sum of the emitted, absorbed, and scattered fluxes appearing on the right-hand side, respectively.

This representation is related to the kinetic theory commonly applied to describe the behavior of a gas of particles as the Radiative Transfer Equation can be derived from the Boltzmann equation. This Radiative Transfer Equation plays a pivotal role in establishing fluxes and temperatures across numerous radiation problems, encompassing situations in both transparent and nontransparent media. A visual representation of this equation is encapsulated in Fig. 1.4, providing an insight into the underlying flux balance that it is expressing.

Emphasizing a pivotal aspect of the conceptual framework employed here is essential, namely, the averaging of the phase of photons. This averaging ensures that the equation cannot directly encapsulate purely wave-like effects, such as

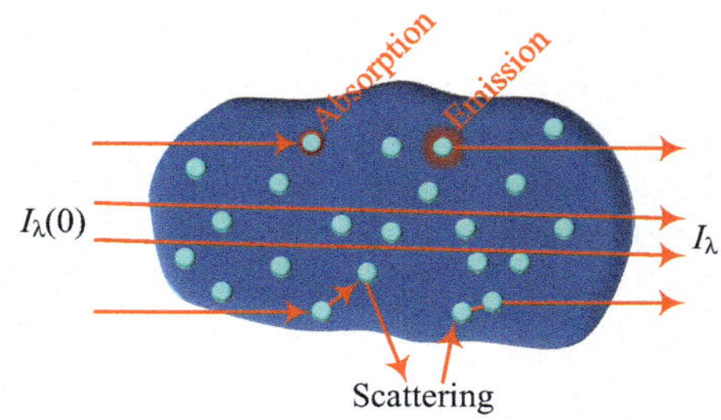

Fig. 1.4 Schematic of the radiative intensity balance in an arbitrary volume element including the mechanisms of emission, absorption, and scattering

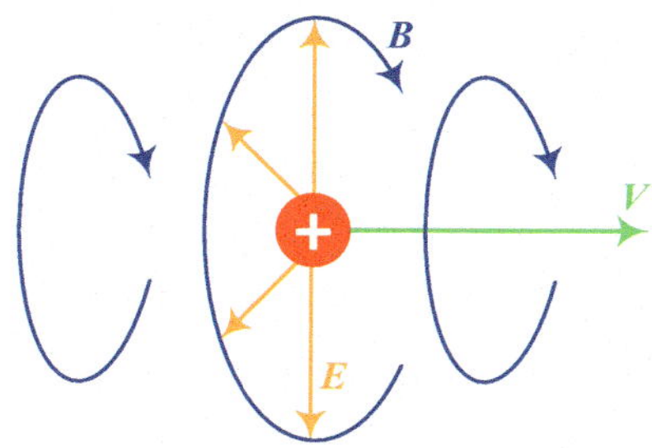

Fig. 1.5 Moving charge producing magnetic and electrical fields. Both fields are decaying with the reverse square of the distance to the charge and have polarizations perpendicular to the charge velocity. One field polarization is also perpendicular to the one of the other field

interferences. Despite this limitation, the Radiative Transfer Equation implicitly addresses a broad range of wave phenomena, including reflection, transmission, absorption, and emission. The implicit inclusion of wave mechanisms within this equation allows for a global understanding and for handling application scenarios at scales ranging from the radiation wavelengths to the one of the universe.

1.2.3 Source of Thermal Radiation

Unraveling thermal radiation takes us deep into the realms of Maxwell's equations, providing an understanding of the source that orchestrates this phenomenon. These equations divulge a fundamental mechanism—any moving charge q sets forth the propagation of an electromagnetic wave defined by the fields, which write, in the nonrelativistic case of a constant charge velocity (Fig. 1.5),

$$\mathbf{E}(r) = \frac{q}{4\pi\epsilon_0 r^2}\hat{r}, \tag{1.4}$$

$$\mathbf{B}(r) = \frac{q\mu_0}{4\pi}\frac{\mathbf{V}\times\hat{r}}{r^2}. \tag{1.5}$$

This effect is the working mechanism of a common antenna used in most wireless communication devices [25]. In the context of solid materials, the charges of ions or electrons, set into motion by thermal agitation, create a propagative electromagnetic disturbance, which gives rise to thermal radiation. Microscopic thermal fluctuations are not accessible to our senses, and they can however be easily demonstrated by the spectrum of the vibrations of a mechanical microresonator such as the cantilever of an atomic force microscope out of contact [26]. Thermal fluctuations are also generating the Brownian motion seen under an optical microscope, which provided the first experimental proof of the existence of the atom [27].

As a consequence, we ascertain that any physical body boasting a nonzero temperature inherently becomes a radiation source. The interpretation lays bare the essence of radiation, fundamentally composed of electromagnetic waves that progress through space. The theoretical underpinnings of this concept find roots in Maxwell's equations, providing a robust framework for comprehending the interplay between thermal agitation, charge movements, electromagnetic field, and thermal radiation as depicted in Fig. 1.6.

The exploration of radiative heat fluxes through the lens of Maxwell's equations is grounded in the definition of the Poynting vector $\langle P \rangle$ at frequency ω and position r, which is generated by the electromagnetic currents $J(r, \omega)$ within the volume of the source:

$$\langle \mathbf{P}_x(\mathbf{r}, \omega) \rangle = i\omega\mu_0 \int_{V'} d^3\mathbf{r}' \int_{V''} d^3\mathbf{r}'' G_{E_x l}(\mathbf{r}, \mathbf{r}') G^{T*}_{H_x m}(\mathbf{r}, \mathbf{r}'') \langle J_l(\mathbf{r}', \omega) J^*_m(\mathbf{r}'', \omega) \rangle, \tag{1.6}$$

where V' and V'' refer to the same source volume described by two different coordinates r' and r''. G is the Green function tensor relating a field at point r to the current at another point, i.e., r' for E and r'' for H. T refers to the matrix transpose, $*$ to the complex conjugate, and μ_0 to the permeability of the empty space. This formula unveils a crucial condition—the necessity to model, in electromagnetic terms, a thermal source, that is to say, the spectrum of the fluctuating currents $\langle J(r', \omega) J^*(r'', \omega) \rangle$ borne by the charges excited by thermal noise within a solid. Pioneering this modeling, Rytov [28] emerged as a trailblazer, employing the Fluctuation–Dissipation Theorem to link current fluctuations to temperature:

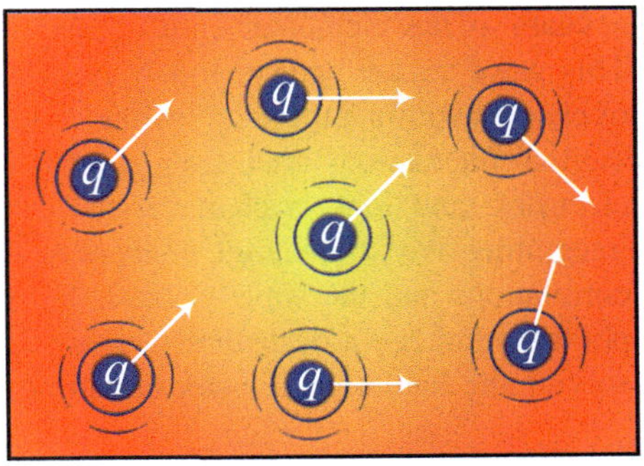

Fig. 1.6 Charge motions due to thermal fluctuations emit thermal radiation. The motions of electrons and ions in a solid are subjected to thermal noise, which yield the electromagnetic field of heat radiation

$$\left\langle J(r', \omega) J^*(r'', \omega) \right\rangle = \frac{\omega \epsilon_0}{\pi} \mathrm{Im}\,(\epsilon(\omega)) \frac{\hbar \omega}{\exp\left(\frac{\hbar \omega}{k_B T}\right) - 1} \delta(r' - r''), \quad (1.7)$$

where $\epsilon(\omega)$ refers to the dielectric function of the medium and ϵ_0 to the dielectric constant of vacuum. The left-hand side term expresses the fluctuating quantities, i.e., the charge current correlation in a given direction, while the imaginary part of the dielectric constant on the right-hand side refers to the electromagnetic absorption in the medium that can be considered as the dissipation process. We also note the presence of the oscillator energy on the same side. Using the expression of the fluctuating currents yields a Poynting vector proportional to $i\omega^2 \mu_0 \epsilon_0 \mathrm{Im}(\epsilon(\omega))$, to the oscillator energy, and also to a space- and frequency-dependent function consisting in a single volume integral of the Green function tensors. This latter is analytically tractable only in simple geometries; otherwise, it has to be computed numerically.

In conclusion, the source of thermal radiation, as deciphered through Maxwell's equations and the groundbreaking work of Rytov, emerges as an interplay between the thermal agitation of charges and the ensuing generation of electromagnetic waves.

1.2.4 Waves, Packets, and Particles

The preceding discourse delves into the nature of radiation, unraveling two distinct perspectives—one grounded in particles, which trajectories are governed by the Radiative Transfer Equation, and the other rooted in the undulating nature of electromagnetic fields. While both viewpoints contribute to our comprehension of radiation, it is the latter, encapsulating the essence of waves, that more profoundly elucidates this phenomenon, preserving essential properties, notably the phase. Setting the conceptual frame at the wave level also provides the distinction between evanescent and guided waves, both of interest in the context of a multiscale description. On a deeper layer of interpretation, the transition between the descriptions provided by Maxwell's equations and the Radiative Transfer Equation finds its foundation in the existence of wave packets, the combinations of electromagnetic plane waves, via the Wigner Transform [29] (Fig. 1.7).

The wave packet picture not only harmonizes the two frameworks but also appears as the elemental entity governing kinetic transport in thermal radiation. This wave packet spontaneously emerges as a correlation of thermal movements of charges within a thermodynamic system at equilibrium. In this equilibrium state, thermal fluctuations arise from stochastic movements of the microscopic entities, electrons, and ions, a characteristic that is averaged through correlation. Upon integrating over the volume element to define local flux, the features of both these correlations and the wave packets disappear, leaving a singular entity—the local photon number—a deterministic statistical distribution of photons provided when solving the Radiative Transfer Equation.

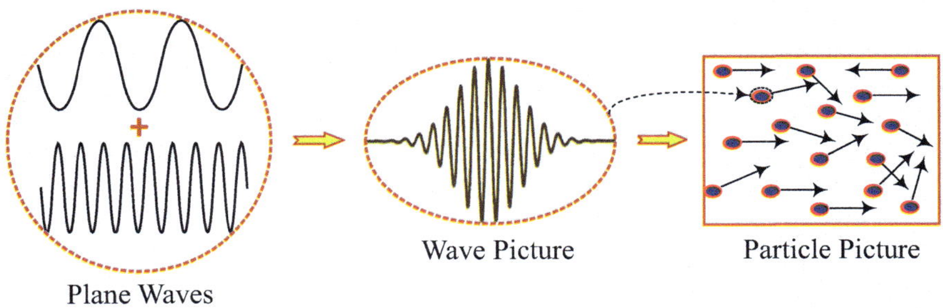

Fig. 1.7 Schematic of the main physical pictures involved in heat radiation. On the left side, the basic electromagnetic modes at play are provided by Maxwell's equations. In the middle, the combination of the basic electromagnetic modes gives rise to wave packets. Wave packets are produced by the stochastic thermal excitation of the modes. Finally, on the right, wave packets are assimilated to particles having trajectories and amplitudes governed by the Radiative Transfer Equation

Emphasizing a crucial nuance, we should clarify here the misconception that the Radiative Transfer Equation deals directly with photon particles. In reality, photons, in this context, are quanta of nonlocal waves. Conversely, wave packets manifest as local and time-dependent entities, bridging the gap between the wave and particle realms. It is paramount to recall that a photon, as the quantum of energy, results from the quantization of the electromagnetic wave. A wave packet, in contrast, combines multiple electromagnetic waves, hinting at the presence of a substantial number of photons within its structure.

1.2.5 Two-Dimensional Gas of Packets

The study of radiation within a solid traditionally relies on the Radiative Transfer Equation applied in the three dimensions of space. This approach, as elucidated earlier, governs the transport of wave packets, charting their trajectories in three dimensions. Correspondingly, the wave packet displacement mirrors the propagation of electromagnetic waves along corresponding directions in space. However, let us now consider that these electromagnetic waves propagate exclusively within a two-dimensional plane. This alteration allows for investigating the electromagnetic waves either from an *ab initio* standpoint, invoking the foundational Maxwell's equations and the Fluctuation–Dissipation Theorem, or by considering the propagation of a gas of wave packets within this two-dimensional space. The eigenmodes of the planar structure mold the packets, and the wave mechanisms operating along the in- and out-of-plane directions are derived from the dispersion relation inherent in the plane geometry using Maxwell equations.

As the planar directions are arbitrarily large, they will not induce wave effects and will behave as usual propagation dimensions. The complex wave problem of propagation in planar assemblies will then be reduced to the trivial kinetics of packets propagating in two dimensions of space, solved by the two-dimensional Radiative Transfer Equation. As we will be tackling surface waves that are propagating within a two-dimensional space, this substantial simplification proposed by G. Chen et al. [15] will constitute the framework of our approaches all along the coming chapters.

1.3 Surface Waves

Surface waves, ubiquitous in our experiential landscape, manifest in various phenomena, ranging from the undulating ocean waves to the seismic ripples of the Earth's crust. In these instances, these waves unfold as gravity [30] and elastic waves [31], respectively. Beyond these terrestrial scenarios, electromagnetic waves are also forming two-dimensional systems (Fig. 1.8).

1.3.1 Propagation Length, Group Velocity, and Amplitude

In a three-dimensional body, a wave of constant amplitude emanating from a volume source and radiating spherically emits a constant flux through the sphere of radius r. The flux density therefore follows an inverse square law ($1/r^2$), diminishing as the radius of the sphere expands. Conversely, the propagation of a surface wave manifests a distinct trajectory—its flux density follows the inverse of a circle length ($1/r$) if r here refers to the radius of a circle. Those two behaviors are illustrated in Fig. 1.9. This distinction is pivotal, yielding a simple consequence: Surface waves propagate over longer distances compared to their volumetric counterparts.

Elastic surface waves are generally slower than bulk waves. This is because the surface is less rigid, since stresses apply only in the material hemisphere and not in all directions of the sphere, as in the body of the material. As we will see, a surface electromagnetic wave is evanescent in the out-of-plane direction, and the component

Fig. 1.8 Examples or surface waves: propagating ocean waves (left) and standing acoustic waves (right) governed by the equations of gravity waves and elasticity, respectively

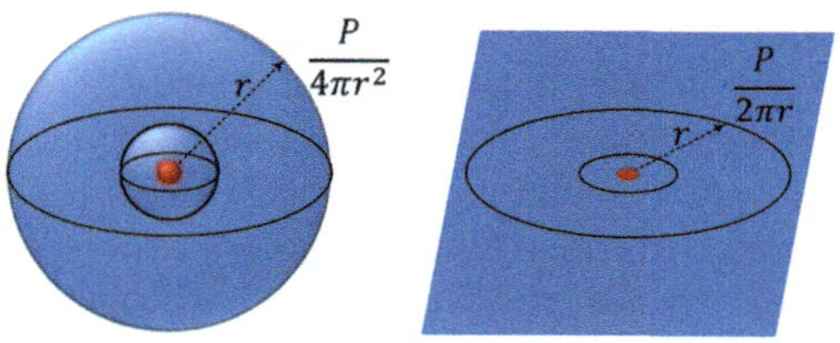

Fig. 1.9 Schematic of the wave front propagation emitted from a point source of power P in three dimensions producing spherical waves (left) and in two dimensions yielding circular waves (right). The power P is conserved when the wave expands, and therefore the powers per unit surface and length follow $1/r^2$ and $1/r$ laws, respectively

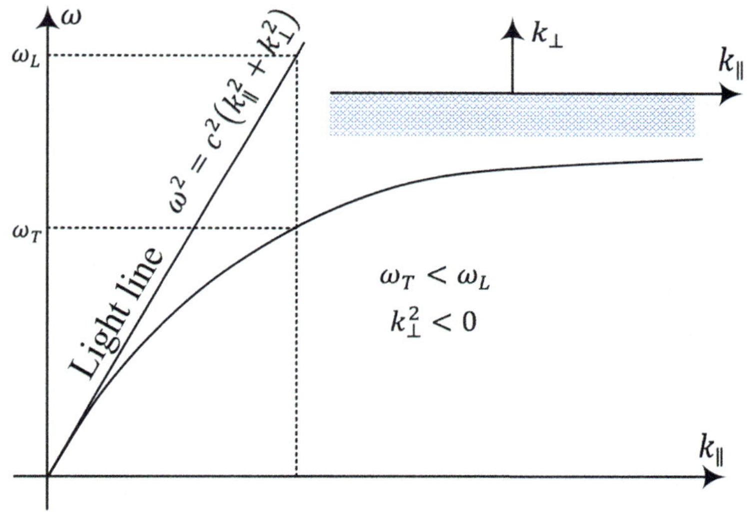

Fig. 1.10 On a surface, the wave vector can be decomposed into components parallel and perpendicular to this surface $\mathbf{k} = \mathbf{k}_\perp + \mathbf{k}_\parallel$. The parallel wave vector refers to the propagating component and has real and imaginary parts, the imaginary one manifesting absorption in the medium $\exp(-\mathrm{Im}(\mathbf{k}_\parallel) \cdot \mathbf{x})$. This parallel wave vector is used to draw the dispersion relations of i) the light in a nondispersive solid featuring a linear dependence between frequency and wave vector $w = ck_\parallel$ (dashed line) and ii) a surface wave that is by nature evanescent in the out-of-plane direction $exp(-k_\perp z)$, i.e., its perpendicular wave vector is a pure imaginary number. The square value of this latter being negative, the corresponding frequency becomes smaller than the one of light and so is its group velocity

of the wave vector in this direction has then to be a pure imaginary number. This condition ensures that the out-of-plane wave amplitude decays exponentially. As depicted in Fig. 1.10, the frequency of this wave will then be lower than that of light in a solid material, and its speed will therefore be also lower than that of the three-dimensional wave.

The kinetic energy of the classical oscillator $k_B T/2 = mv^2/2$ indicates that the velocity v of an atom of mass m is proportional to the square root of the temperature. In a polar material, this movement induces the agitation of charges or dipoles and a corresponding current correlation $\langle qv(r) \cdot qv(r') \rangle$ proportional to temperature. The emitted power of an electromagnetic mode related to such a phonon–polariton excitation should therefore be proportional to T in the classical limit. At lower temperatures ($< \hbar\omega/k_B$), this power has to be set proportional to the oscillator energy, i.e., the product between the energy quantum $\hbar\omega$ and the Bose-Einstein distribution $(\exp(\hbar\omega/k_B T) - 1)^{-1}$, which is the number of energy quanta on the mode. The amplitude of the mode clearly appears as equal to the number of energy quanta that it is carrying. The energy of a plasmon mode has therefore the same dependence to temperature and frequency. In the case of a body at thermal equilibrium, the temperature being the same at the surface and inside the body, the polariton amplitudes are the same at the surface and in the core of the material.

1.3.2 Surface and Bulk Waves

The equations governing wave propagation in the bulk material, such as the elasticity equation or the electromagnetic propagation equation, are generally simpler than those for surface waves, as the latter must take into account the complexity of interface boundary conditions involving two media. In bulk materials, the electric field propagation equation leads directly to the dispersion equation $k = (\omega/c)\sqrt{\epsilon}$ reported in Fig. 1.11 [32], linking the pulsation ω to the wave vector k with the speed of light c and the dielectric function ϵ. The latter expresses the coupling between charges and the electromagnetic field. In particular, polariton modes manifest themselves across a range of negative values of the real part of ϵ, where a band gap divides the light line into two branches (Fig. 1.11).

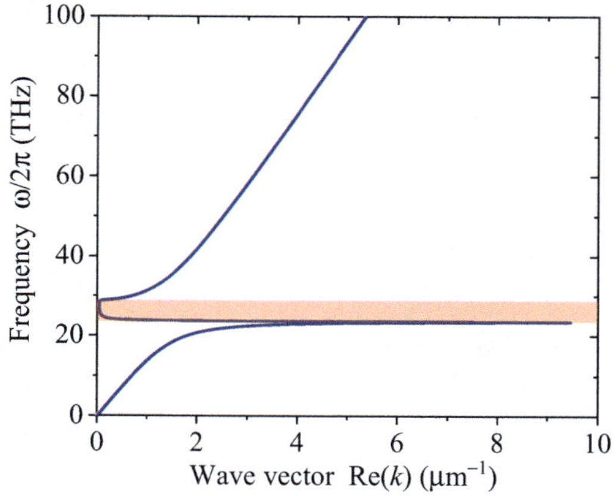

Fig. 1.11 Dispersion relation of bulk phonon–polaritons highlighting a band gap (reddish zone) in the frequency range where the dielectric function is negative. The dielectric function is a Lorentzian including a single resonance, typically referring to polar materials such as SiC

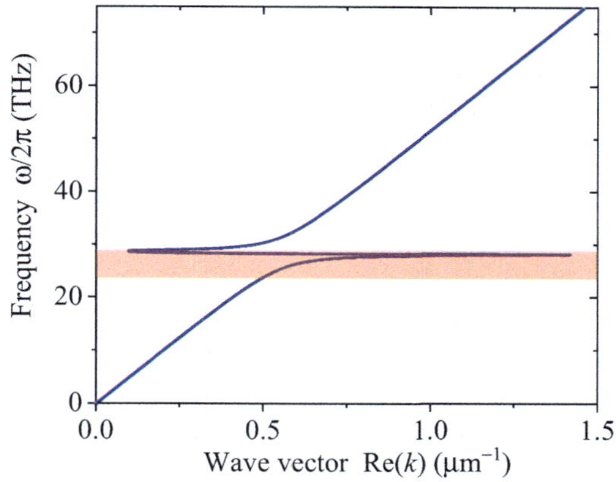

Fig. 1.12 Dispersion relation of a surface phonon–polariton highlighting a band referred to as the "Reststrahlen band" (reddish zone), including an additional peak due to absorption in the medium"

We will show later that the introduction of a surface with vacuum or air radically modifies this latter dispersion into $k = (\omega/c)\sqrt{\epsilon/(\epsilon+1)}$ inducing a smaller group velocity as shown in Fig. 1.12 [2].

References

1. Fano, U.: Atomic theory of electromagnetic interactions in dense materials. Phys. Rev. **103**, 1202 (1956)
2. Agranovich, V.M.: Surface Polaritons, vol. 1. North Holland, Amsterdam (2012)
3. Pan, Z., Lu, G., Li, X., McBride, J.R., Juneja, R., Long, M., Lindsay, L., Caldwell, J.D., Li, D.: Remarkable heat conduction mediated by non-equilibrium phonon polaritons. Nature **623**, 307 (2023)
4. Tervo, E.J., Boyuk, D.S., Cola, B.A., Zhang, Z.M., Filler, M.A.: Sub-diffractional waveguiding by mid-infrared plasmonic resonators in semiconductor nanowires. Nanoscale **10**, 5708 (2018)
5. Tervo, E.J., Gustafson, M.E., Zhang, Z.M., Cola, B.A., Filler, M.A.: Photonic thermal conduction by infrared plasmonic resonators in semiconductor nanowires. Appl. Phys. Lett. **114**, 163104 (2019)
6. Polder, D., Van Hove, M.: Theory of radiative heat transfer between closely spaced bodies. Phys. Rev. B **4**, 3303–3314 (1971)
7. Loomis J.J., Maris, H.J.: Theory of heat transfer by evanescent electromagnetic waves. Phys. Rev. B **50**, 18517–18524 (1994)
8. Joulain, K., Mulet, J.-P., Marquier, F., Carminati, R., Greffet, J.-J.: Surface electromagnetic waves thermally excited: radiative heat transfer, coherence properties and Casimir forces revisited in the near field. Surf. Sci. Rep. **57**, 59–112 (2005)
9. Rousseau, E. Siria, A., Jourdan, G., Volz, S., Comin, F., Chevrier, C., Greffet, J.-J.: Radiative heat transfer at the nanoscale. Nat. Photon. **3**, 514–517 (2009)
10. Shen, S., Narayanaswamy, A., Chen, G.: Surface phonon polaritons mediated energy transfer between nanoscale gaps. Nano Lett. **9**, 2909–2913 (2009)
11. Song, B., Thompson, D., Fiorino, D., Ganjeh, Y., Reddy, P., Meyhofer, E.: Radiative heat conductances between dielectric and metallic parallel plates with nanoscale gaps. Nat. Nanotechnol. **11**, 509 (2016)
12. Laroche, M., Carminati, R., Greffet, J.-J.: Near-field thermophotovoltaic energy conversion. J. Appl. Phys. **100**, 063704 (2006)

13. Lucchesi, C., Cakiroglu, D., Perez, J.-P., Taliercio, T., Tournié, E., Chapuis, P.-O., Vaillon, V.: Near-field thermophotovoltaic conversion with high electrical power density and cell efficiency above 14%. Nano Lett. **21**, 4524–4529 (2021)
14. Greffet, J.-J., Carminati, R., Joulain, K., Mulet, J.-P., Mainguy, S., Chen, Y.: Coherent emission of light by thermal sources. Nature **416**, 61–64 (2002)
15. Chen, S.A., Narayanaswamy, A., Chen, G.: Surface phonon-polariton mediated thermal conductivity enhancement of amorphous thin films. Phys. Rev. B **72**, 155435 (2005)
16. Tranchant, L., Hamamura, S., Ordonez-Miranda, J., Yabuki, T., Vega-Flick, A., Cervantes-Alvarez, F., Alvarado-Gil, J.J., Volz, S., Miyazaki, K.: Two-dimensional phonon-polariton heat transport. Nano Lett. **19**, 6924 (2019)
17. Wu, Y., Ordonez-Miranda, J., Gluchko, S., Anufriev, R., De Sousa Meneses, D., Del Campo, L., Volz, S., Nomura, M.: Enhanced thermal conduction by surface phonon-polaritons. Sci. Adv. **6**, eabb4461 (2020)
18. Ordonez-Miranda, J., Tranchant, L., Kim, B., Chalopin, Y., Antoni, T., Volz, S.: Quantized thermal conductance of nanowires at room temperature due to Zenneck surface-phonon polaritons. Phys. Rev. Lett. **112**, 055901 (2014)
19. Pei, Y., Chen, L., Jeon, W., Liu, Z., Chen, R.: Low-dimensional heat conduction in surface phonon polariton waveguide. Nat. Commun. **14**, 8242 (2023)
20. Salihoglu, H., Iyer, V., Taniguchi, T., Watanabe, K, Ye, P., Xu, X.: Energy transport by radiation in hyperbolic material comparable to conduction. Adv. Funct. Mater. **30**(6), 1905830 (2019)
21. Chen, Y., Segovia Pacheco, M.A., Salihoglu, H., and Xu, X.: Greatly enhanced radiative transfer enabled by hyperbolic phonon polaritons in $\alpha-MoO_3$. Adv. Funct. Mater. 2403719 (2024)
22. Kittel, C.: Introduction to Solid State Physics. Wiley, London (2004)
23. Howell, J.R., Pinar Menguc, M., Daun, K.: Thermal Radiation: An Introduction. CRC Press, Boca Raton (2023)
24. Planck, M.: Theory of Heat Radiation. Berlin (1914)
25. Balanis, C. A.: Antenna Theory: Analysis and Design. Wiley, London (2016)
26. Ma, C., Zhou, C. Peng, P., Chen, Y., Arnold, W., Chu, J.: Thermal noise in contact atomic force microscopy. J. Appl. Phys. **129**, 234303 (2021)
27. Perrin, J.: Brownian Movement and Molecular Reality. Taylor and Francis, London (1910)
28. Rytov, S.M.: Theory of Electric Fluctuations and Thermal Radiation. Science Academy Press (1953)
29. Doicu, A., Mishchenko, M.I.: An overview of methods for deriving the radiative transfer theory from the Maxwell equations. III: effects of random rough boundaries and packing density. J. Quant. Spectrosc. Radiat. Transfer **224**, 154–170 (2019)
30. Nappo, C.J.: An Introduction to Atmospheric Gravity Waves. Academic Press, New York (2012).
31. Royer, D., Valier-Brasier, T.: Elastic Waves in Solids 1: Propagation. Wiley, London (2022)
32. Dresselhaus, M.S.: Solid State Physics, Part II, Optical Properties of Solids. Online

Chapter 2
Fundamentals of Thermal Radiation Driven by Polaritons

2.1 Heat Radiation Theory

2.1.1 Three-Dimensional Planck and Stefan–Boltzmann's Laws

Planck's Law [1], formulated at the outset of the 20th century predating the advent of quantum mechanics, stands as a foundational equation. It has persisted as a reference equation extensively employed in the fields of thermal radiation and spectroscopy.

Let us begin by considering a volume element residing within either vacuum or air—two mediums with complete transparency and possessing unitary dielectric functions. This volume element is surrounded by isothermal walls at temperature T. Within this selected volume, electromagnetic waves traverse in all directions of the sphere. The focus of our examination lies in the energy encapsulated within the spectral range spanning from frequency f to $f + df$, where df is a frequency differential, within this volumetric unit. This energy is the product between the energy of the electromagnetic mode at a given frequency and the number of modes within the designated frequency interval (Fig. 2.1).

In the quantum framework, the energy of a mode can be expressed as the product of the energy quantum (hf) and the number of energy quanta associated with that mode, determined by the Bose–Einstein distribution $n^0(f, T) = 1/\left(\exp(hf/k_B T) - 1\right)$. Within the frequency interval df, the total number of modes is defined by the quantity called density of states, denoted as $g(f)$. Additionally, in the derivation of the vacuum energy expression, it is imperative to factor in the number of electromagnetic polarizations, acknowledging their count of two, yielding the final expression for the energy per unit volume and at frequency f, u_f:

$$u_f = 2hf n^0(f, T) g(f). \tag{2.1}$$

S. Volz, J. Ordonez-Miranda, *Heat Transport Driven by Surface Electromagnetic Waves*, Mechanical Engineering Series, https://doi.org/10.1007/978-3-031-66024-5_2

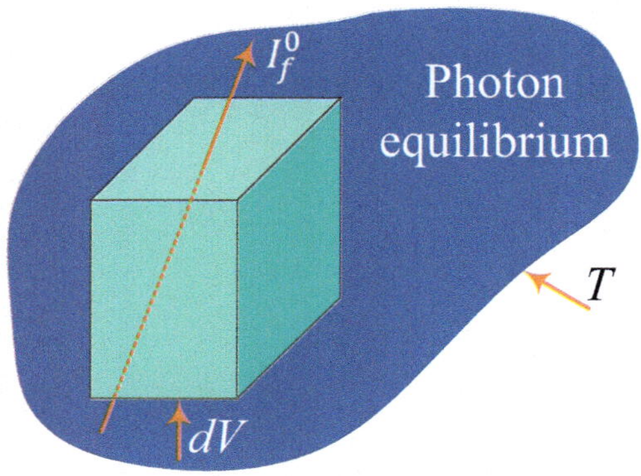

Fig. 2.1 Scheme of a volume element dV of vacuum or air crossed by electromagnetic waves in various directions. This volume element is assumed to be surrounded by isothermal walls at temperature T yielding a bath of photons at thermal equilibrium inside the cavity. The radiative intensity I_f^0 (the spectral heat flux emitted in one direction, crossing the volume element) is provided by Planck's Law

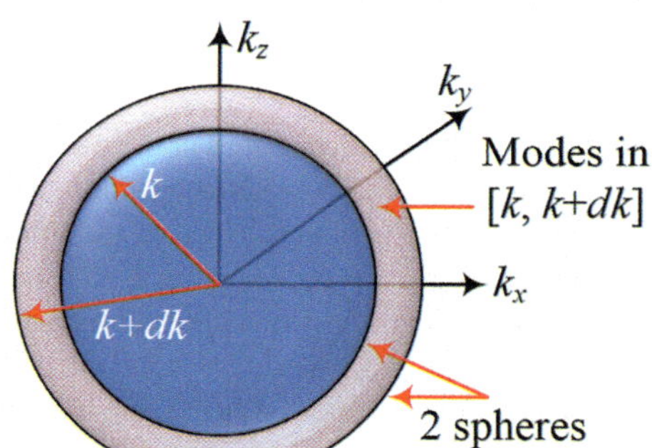

Fig. 2.2 Three-dimensional reciprocal space where the sub-volume comprised between spheres of radii k and $k + dk$ is depicted. This volume encompasses the modes with frequencies ranging between f and $f + df$ when noticing that $k = 2\pi f/c$

The pivotal factor in this expression is the density of states, denoted as $g(f)$, which is determined through a process akin to that used for electrons and phonons. The derivation involves enumerating the states in the three-dimensional reciprocal space (k_x, k_y, k_z). This procedure entails assessing the volume enclosed in the $[k, k + dk]$ interval, i.e., by spheres with radii k and $k+dk$. This volume is defined by the sphere surface multiplied by the radii difference dk as $4\pi k^2 dk$, where $k = \sqrt{k_x^2 + k_y^2 + k_z^2}$ represents the modulus of the wave vector. To determine the number of states in this latter volume, the assessment of the volume element occupied by a single state becomes necessary (Fig. 2.2).

In this objective, we initiate the process in a single dimension, focusing on a plane wave constrained by a period defined by an arbitrary length L. This length is set to the system size which is to be duplicated via a translation to form a periodic set of identical systems. The periodic condition, often referred to as the Born–Von-Karman boundary condition, dictates that $\exp[ik.(r + L)] = \exp(ik.r)$, which yields $\exp(ikL) = 1$. This expression further implies that $k = 2\pi n/L$, with n being an integer. The wave vectors are evidently discretized, with the interval between two consecutive values defined as $2\pi/L$, representing the segment occupied by a single wave vector or mode. Consequently, the volume element occupied by one mode in three dimensions is ultimately given by $(2\pi/L)^3$. Noting that $k = 2\pi f/c$, the density of states per unit volume of the real space is derived from the volume in $[k, k + dk]$, i.e., $4\pi k^2 dk$, divided by the volume element of one mode $(2\pi/L)^3$ as:

$$g(f) = \frac{4\pi f^2}{c^3}. \tag{2.2}$$

Note that the number of modes per unit volume in the $[f, f + df]$ interval is derived from the product between this density of states and the frequency interval df. The energy density (J.m^{-2}.Hz^{-1}) in (2.1) thus takes the form:

$$u_f = \frac{8\pi h f^3}{c^3} n^0(f, T). \tag{2.3}$$

The radiative intensity I_f defined as the heat flux in a given direction is derived by dividing this energy by 4π, the solid angle of the sphere, and multiplying it by the light speed:

$$I_f = \frac{2hf^3}{c^2} n^0(f, T). \tag{2.4}$$

The intensity (W.m^{-2}sr^{-1}.Hz^{-1}) provided by this Planck's Law is independent of the direction and relies solely on frequency and temperature through n^0.

2.1.1.1 Stefan–Boltzmann's Law

Conventional engineering approaches to characterize intricate systems and various heat transfer mechanisms often necessitate a straightforward formulation for the flux emitted by a surface. One solution involves integrating Planck's intensity across the spectrum, leading to the derivation of the following expression:

$$\int_0^{\infty} I_f^0(T)\, df = \frac{2\pi^4 k_B^4}{15c^2 h^3} \cdot T^4 = \frac{\sigma}{\pi} T^4 = I^0, \tag{2.5}$$

where $\sigma = 2\pi^5 k_B^4/(15c^2h^3) \approx 5.67 \times 10^{-8}$ W.m^{-2}.K^{-4} is the Stefan–Boltzmann constant. This flux I^0 (W.m^{-2}.sr^{-1}) is directly proportional to the temperature to the 4th power. The total flux Φ_{dS} per unit surface emitted by a Blackbody surface is then obtained by integrating the latter expression over the hemisphere:

$$\Phi_{dS} = \frac{1}{dS} \int_{\Omega} I^0\, dS \cos\theta\, d\Omega, \tag{2.6}$$

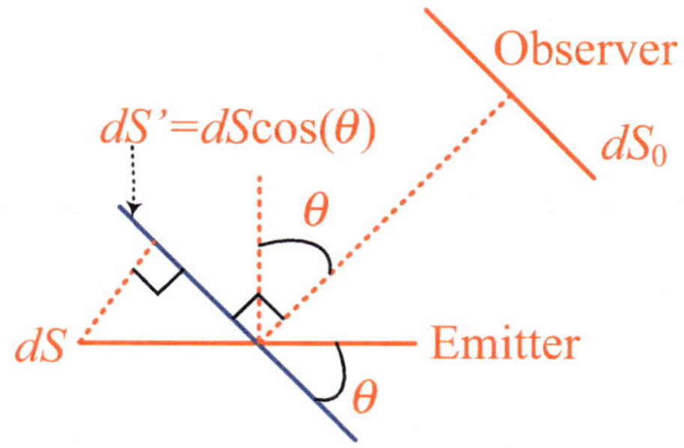

Fig. 2.3 The emitted power from the surface element dS to dS_0 requires the projection of the surface dS into dS_0 plane via the cosine of the angle between the normals of dS_0 and dS

where dS is the emitting surface element. Considering the reception surface dS_0, the flux received from dS in a given angle reads $I^0 dS \cos\theta d\theta$ as explained in Fig. 2.3.

Considering rotational symmetry around the normal to dS, the solid angle element $d\Omega = 2\pi \sin(\theta)d\theta$ leads to

$$\Phi_{dS} = 2\pi \int_{\Omega} I^0 \cos\theta \sin\theta \, d\theta. \tag{2.7}$$

As I^0 does not depend on angle, one obtains

$$\Phi_{dS} = 2\sigma \cdot T^4 \int_0^{\frac{\pi}{2}} \sin\theta \cos\theta \, d\theta = \sigma T^4. \tag{2.8}$$

This expression corresponds to the well-known Stefan–Boltzmann's Law giving the total heat flux per unit surface (W.m^{-2}) emitted by a blackbody.

2.1.2 *Radiative Transfer Equation and Boltzmann Equation*

In Chap. 1, the Radiative Transfer Equation was provided in this form:

$$\nabla I_\lambda = \epsilon_\lambda (I_\lambda^0 - I_\lambda) + \sigma_\lambda \left[\langle I_\lambda \rangle_\Omega - I_\lambda \right], \tag{2.9}$$

where ϵ_λ is the emissivity and σ_λ the scattering coefficient. Note that in the forthcoming chapters, the systems under examination will remain homogeneous, thus making it safe to eliminate the scattering term. We now consider the intensity as a quantified heat flux [2] $I_{f,\Omega}(r) = 2hf n_f(r) v_{f,\Omega}$, where the factor 2 arises from the two polarizations, and $v_{f,\Omega}$ refers to the mode group velocity. This velocity may differ from the one of light as the volume element might be at the core of a material with a non-unitary dielectric function. The number n_f refers to the local and averaged number of photons per unit volume on the mode. With this knowledge in mind, the aforementioned relation can be reformulated as

$$\mathbf{v}_{f,\Omega} \cdot \nabla n_f(\mathbf{r}) = v_{f,\Omega}\epsilon_f \left[n_f^0(\mathbf{r}) - n_f(\mathbf{r})\right], \tag{2.10}$$

where a mode relaxation time can be identified as $\tau_f = \left(v_{f,\Omega}\epsilon_f\right)^{-1}$. The well-known Boltzmann Transport Equation written in the single mode relaxation time approximation arises as follows [3, 4]:

$$\mathbf{v}_{f,\Omega} \cdot \nabla n_f(\mathbf{r}) = \frac{n_f^0(\mathbf{r}) - n_f(\mathbf{r})}{\tau_f}, \tag{2.11}$$

where n_f^0 denotes the photon number at equilibrium or the Bose–Einstein distribution. This expression serves as the foundation for deriving the phonon thermal conductivity, which we will utilize to now estimate the radiative thermal conductivity.

2.1.3 *Radiative Thermal Conductivity in Bulk Materials*

The radiative or Rosseland thermal conductivity [5] is derived in a three-dimensional body locally submitted to a unidirectional temperature gradient. The medium is supposed to be sufficiently absorbing such that the volume element is larger than the attenuation lengths. The total flux along the temperature gradient direction can be established from: i) the projection of the radiative intensities onto this same direction and ii) their frequency integration, as:

$$\phi(\mathbf{r}) = \int_f^{\infty} \int_{\theta=-\frac{\pi}{2}}^{\frac{\pi}{2}} I_f(\mathbf{r}) \cos\theta \, d\Omega \, df = 4\pi \int_0^{\infty} \frac{hf^3}{v^2} \int_{\theta=-\frac{\pi}{2}}^{\frac{\pi}{2}} n_f(\mathbf{r}) \cos\theta \sin\theta \, df, \tag{2.12}$$

where the velocity v_f in the medium is assumed to be isotropic, $g(f)$ denotes the density of states, and θ refers to the angle between the direction of the flux intensity and the one of the temperature gradient (x). $n_f(r)$ corresponds to the nonequilibrium photon number in the medium and is provided by the Boltzmann transport equation, as follows:

$$n_f(\mathbf{r}) = n_f^0(\mathbf{r}) - v_f \tau_f \cos\theta \frac{\partial T(x)}{\partial x} \frac{\partial n_f^0}{\partial T}. \tag{2.13}$$

The spatial gradient was decomposed into a composition of derivatives, and the derivative of the nonequilibrium photon number was also considered equal to the derivative of the equilibrium one, since we mentioned that the dimensions of the volume element remain smaller than the decay length of the modes. Combining the above and the heat flux expressions leads to

$$\phi(\mathbf{r}) = -4\pi \int_0^\infty \frac{hf}{v_f^2} \frac{\partial n_f^0}{\partial T} f^2 \Lambda \, df \int_{\theta=-\frac{\pi}{2}}^{\frac{\pi}{2}} \cos^2\theta \sin\theta \, d\theta \cdot \frac{\partial T(x)}{\partial x},$$
$$- 4\pi \int_0^\infty \frac{hf}{v_f^2} n_f^0(\mathbf{r}) f^2 \, df \int_{\theta=-\frac{\pi}{2}}^{\frac{\pi}{2}} \cos\theta \sin\theta \, d\theta. \tag{2.14}$$

The second term, being the heat flux at equilibrium, cancels, and a radiative thermal conductivity $\kappa = -\phi(\mathbf{r})/(\partial T/\partial x)$ can be defined through the Fourier's Law, which yields

$$\kappa = \frac{2}{3\pi} \int_0^\infty hf \Lambda k^2 \frac{\partial n_f^0}{\partial T} \, df, \tag{2.15}$$

where $\Lambda = v_f \tau_f$ is the attenuation length of the mode. This length might also be viewed as an absorption length in the radiative heat transfer terminology. We, however, keep the vocabulary of electromagnetism as Maxwell's equations will be used to derive this quantity. It is indeed remarkable that the above thermal conductivity can be obtained from the information of the sole dispersion relation relating the frequency to the complex wave vector, since $\Lambda = [2\mathrm{Im}(k)]^{-1}$. The dispersion relation $k(f)$ can readily be derived from Maxwell equations in a homogeneous medium.

2.1.4 2D Planck Function and Stefan–Boltzmann Law

When considering the propagation of waves along surfaces and films, we will adopt a similar approach than above but within a two-dimensional space. Accordingly, we define the energy in this specific geometry using $u_f = 2hfn_f^0 g(f)$ and then proceed with calculating the density of states g(f) in a reciprocal space with only two dimensions. As illustrated in Fig. 2.4, the density of states per unit surface is determined by the product of the surface of the reciprocal space enclosed between disks with radii $k_\parallel$ and $k_\parallel + dk_\parallel$ on one hand, and the number of modes per unit surface on the other hand. The subscript $\parallel$ denotes the in-plane directions. The former term can straightforwardly be written as $2\pi k_\parallel dk_\parallel$, while the latter, following the same reasoning as for the three-dimensional case, can be set to $(L/2\pi)^2$, with L being the system size. The resulting density of states $g(f) = \frac{2\pi f}{c^2}$ and the energy density in the frequency interval $[f, f + df]$ then take the form:

$$u_f = \frac{4\pi h f^2 n^0}{c^2}, \tag{2.16}$$

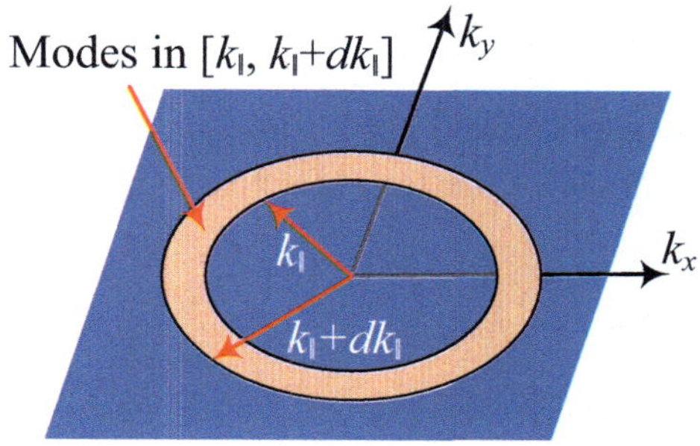

Fig. 2.4 Scheme illustrating the estimation of the surface element in the reciprocal space where modes have vectors in the interval $[k_{\|}, k_{\|}+dk_{\|}]$ or $[f, f+df]$

in units of J.m^{-2}.Hz^{-1}. The intensity in one direction of the plane is obtained by multiplying the energy by the light velocity and dividing it by the directions of the circle, 2π:

$$I_f^0 = \frac{2hf^2n^0}{c}. \tag{2.17}$$

The units of this two-dimensional (2D) flux are W.m^{-1}.sr^{-1}.Hz^{-1}. This intensity establishes the two-dimensional Planck's Law. It is not angle dependent but projecting its contributions coming from the directions of the half circle, onto a single direction requires introducing the cosine of the azimuthal angle ϕ between the rays and the chosen projection direction. As $\int_{-\frac{\pi}{2}}^{\frac{\pi}{2}} \cos(\phi)\, d\phi = 2$, the integration over angles yields another factor of 2 in the resulting hemicylindrical flux $4\frac{hf^2n^0}{c}$. Integrating this latter flux over frequency yields a flux (W.m^{-1}) per unit thickness element db as

$$\phi_{\mathrm{db}} = \sigma_{\mathrm{2D}} T^3, \tag{2.18}$$

where the 2D Stefan–Boltzmann constant is given by [6]

$$\sigma_{\mathrm{2D}} = \frac{4\zeta(3)k_B^3}{ch^2} = 9.61 \times 10^{-11}\,\mathrm{W.m^{-1}.K^{-3}}, \tag{2.19}$$

with $\zeta(3) \approx 1.2021$ being the Riemann zeta function. The flux in W.m^{-2} can be derived from the emission temperature and the thickness of the film b as $\sigma_{\mathrm{2D}}T^3/b$. As a comparison, a one-micron thick film at 300K emits 2596 W.m^{-2}, which is about six times the flux predicted by the standard 3D Stefan–Boltzmann's Law. This astonishing result highlights the fact that the 2D density of states is larger than the 3D one when considering thin enough systems. The ratio between the 2D and 3D fluxes reads $\sigma_{\mathrm{2D}}/(\sigma_{\mathrm{3D}}Tb)$ and shows the predominance of the usual 3D radiation at high temperatures and for thick bodies.

2.1.5 Radiative Thermal Conductivity in Two Dimensions

A 2D thermal conductivity can also be defined in films by following the above definition of the total heat flux:

$$\phi(r) = \int_0^{\infty} \int_{\phi=0}^{2\pi} I_f(r)\, df \, \cos\phi \, d\phi = \int_0^{\infty} \frac{hf \cdot f}{v_f} \int_{\phi=0}^{2\pi} n_f(r) \, \cos\phi \, d\phi \, df, \tag{2.20}$$

where the expression of I_f is modified to involve only one polarization. The angular integration on the azimuth ϕ serves as a projection of all intensities onto the propagation direction. The final expression for the 2D radiative thermal conductivity for one polarization reads

$$\kappa = \frac{1}{2b} \int_0^{\infty} hf \Lambda k_{\|} \frac{\partial n_f^0}{\partial T} \, df, \tag{2.21}$$

where b again denotes the film thickness. The formula in (2.21) was also obtained in the foundational paper [7], by integrating over angular frequency. Let us remind that this expression only applies in absorbing films where the surface element is larger than the attenuation length. We would also like to remind again that this conductivity can be calculated from the sole knowledge of the dispersion relation providing $k_{\|}$ and Λ as a function of the frequency. Chapters 3 and 4 will develop in detail the derivations of dispersion relations in various planar geometries allowing for the estimation of their thermal properties. As we will see in the next chapters, the polarizations in the above formula are not the TM and TE ones that are present in vacuum and that may also be present in a three-dimensional system. It will be proven that a film actually carries four modes, one symmetric and one antisymmetric on each polarization TE and TM. However, the TE modes have small propagation lengths and are negligibly contributing to the heat flux, and only two TM modes remain. Among those TM modes, one might even be neglected toward the other for small thicknesses.

2.1.6 Radiation in One Dimension

The wire (1D) geometry may also be treated in the same terms when noticing that the density of states writes $g(f) = \frac{1}{c}$ and the energy per unit length arises as $u_f = 2hfn^0/c$. The intensity turns out as $I_f = 2hf \cdot n^0$, and the 1D Planck's Law comes out as

$$\phi_{1D} = \sigma_{1D} T^2, \tag{2.22}$$

where $\sigma_{1D} = \pi^2 k^2/3h \approx 9.46 \times 10^{-13}$ W.K^{-2}. Consequently, a monodimensional waveguide of radiative heat emits 8.52×10^{-8} W at room temperature, whatever its cross section. A waveguide with a cross section of 1μm^2 would thus emit the considerable amount of 85.2 kW.m^{-2}, which is 185 times the flux of a 3D emitter. This impressive outcome is once again related to the difference between the one- and three-dimensional density of states. Forming a waveguide of thermal rays is however not an easy task, as a strong coherence should be maintained all along and inside the linear structure. Another challenge is to make sure that the thermal rays stay parallel in air as numerical aperture or diffraction always tends to broaden the beam and convert it into a three-dimensional one.

Finally, following the previous courses of calculation, the 1D thermal conductivity can also be established, for one polarization, as:

$$\kappa = \frac{1}{ab} \int_0^\infty hf \Lambda \frac{dn_f^0}{dT} df, \tag{2.23}$$

where the product ab corresponds to the cross section of the one-dimensional thermal waveguide. The attenuation Λ is obtained from the dispersion relation and Maxwell equations. If this attenuation length becomes larger than the system length ($\Lambda \gg L$), then the system length becomes the characteristic length, and the thermal conductance κ/L can be written:

$$\frac{\kappa}{L} = \frac{g_0}{ab}, \tag{2.24}$$

where the universal quantum of thermal conductance [8] $g_0 = \frac{k^2T}{h} \int_0^\infty \frac{x^2 e^x}{(e^x-1)^2} dx = \frac{\pi^2 k^2 T}{3h}$, with $x = hf/kT$. This quantum of conductance $g_0 = 9.46 \times 10^{-13} T$ (W.K^{-1}) does not depend on the material properties but is proportional to temperature. The apparition of the quantum of conductance requires a one-dimensional and ballistic propagation. The integral over x also has to converge to its limit, i.e., its integrand should tend to zero within the spectral range of thermal radiation. In the case of silicon carbide or silicon dioxide, this latter condition is met in most situations for temperatures below 2000 K. It is finally remarkable that $g_0 = \sigma_{1D} T$ due to a factor of 2 in the Planck integrals over x. As vacuum and medium velocities are not involved in the one-dimensional case, the flux in a lossless emitter equals the one in vacuum.

2.2 Polaritons

2.2.1 Bulk Polariton Dispersion

2.2.1.1 Dielectric Function of Polar Materials and Phonon–Polaritons

The dispersion relations of the electromagnetic waves associated to a polaritonic resonance in one or two dimensions appeared as the pivotal information to derive their radiative properties in those geometries. In the bulk material, two methods can be followed to extract those dispersions, the first is based on the coupling between the motion equations of the charges and the Maxwell equations [9], and the second consists in establishing the dielectric function and deriving the dispersion of a homogeneous system.

Following the second method, we define the dielectric function by considering a polar diatomic crystal with lattice cells including positive and negative charges ($\pm e$). The charges are associated to interacting masses, supposed to take the same value m, and the harmonic or spring force relating them is characterized by a force constant κ. The motion equation of the ions provides their displacement r as a function of the electric force and field, which yields in turn the expression of the polarization $P = Ner$, where N refers to the volume density of the ionic charges, as a function of the same field. The dielectric function directly comes out as the proportionality coefficient between polarization and electric field. The detailed derivation of the dielectric function will be proposed in Chap. 3 and finally stands as

$$\varepsilon = \varepsilon_\infty + \frac{Ne^2}{\varepsilon_0 m(\omega_T^2 - \omega^2)}, \tag{2.25}$$

where we have introduced the optical phonon frequency $\omega_T = \sqrt{\kappa/m}$ and the vacuum permittivity ε_0. Let us remind here that the material permittivity is nothing else than the product of the dielectric function by the vacuum permittivity. In this derivation, the electronic polarization is neglected as the dipoles in consideration are related to the interaction between ions. Those ion dipoles have oscillations in the TeraHertz range corresponding to the thermal InfraRed spectrum, while the electron-core dipole has resonances at higher frequencies ranging from 100 to 1000 THz.

As illustrated in Fig. 2.5, this dielectric function manifests the presence of a single pole or resonance such as in silicon carbide.

In realistic materials with more than two atoms per cell, several resonances might appear and a fitting function including a sum of poles is then applied. It is notable that the dielectric function takes negative values in a limited frequency range, which expresses a radical change in polarization sign or phase. The imaginary dielectric function of silicon carbide also reveals a resonance at the frequency where the real part is canceling. As the imaginary part of the dielectric constant is related to dissipation, strong absorption and reduced propagation length are expected at

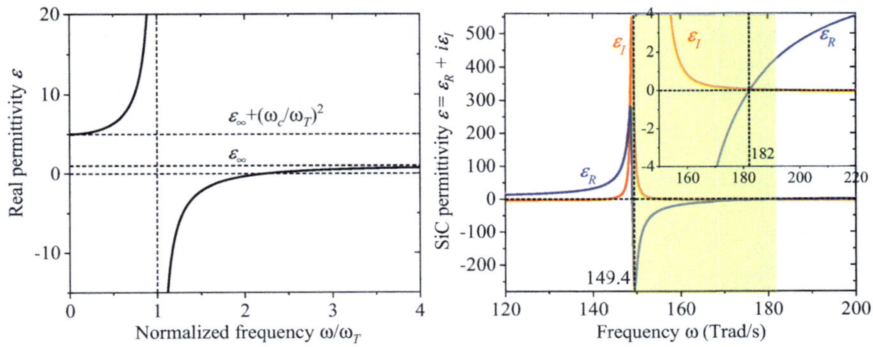

Fig. 2.5 (left) Frequency spectrum of the real dielectric function predicted by the atomic dipole model with $\varepsilon_\infty = 1$, $\omega_c = Ne^2/m\varepsilon_0 = 2\omega_T$. (right) Real and imaginary parts of the dielectric function of silicon carbide as a function of angular frequency

the phonon–polariton resonance. The thermal conductivity formula derived in this chapter then confirms that the contribution of the resonance to heat transfer is not predominant. The above derivation of the dielectric function does not take into account absorption. This effect can be tackled by introducing a damping term in the ion motion equation in the form of a time derivative of the position, which translates into an imaginary term. The resulting dielectric function, keeping $\varepsilon_\infty = 1$:

$$\varepsilon = 1 + \frac{Ne^2}{\epsilon_0 m(\omega_T^2 - \omega^2 - i\Gamma\omega)}, \tag{2.26}$$

now includes an imaginary part that manifests the absorption of the electromagnetic energy in the medium.

2.2.1.2 Dielectric Function of Metals and Plasmon–Polaritons

As explained in Chap. 1, in the case of metals, the moving electron gas is described as superimposed to the ion crystal to produce the plasmon dipolar effect. As a consequence, only the force on electrons should be taken into account in the above motion equation, and the ω_T term, corresponding to ionic forces, should be discarded. The resulting dielectric function is a correction of the one in polar material [9]:

$$\varepsilon = 1 - \frac{n_e e^2}{\epsilon_0 m(\omega^2 + i\Gamma\omega)} \tag{2.27}$$

and is known as the Drude model, which will also be thoroughly derived in Chap. 3. n_e refers here to the electron number density, m to the electron mass, and the plasma

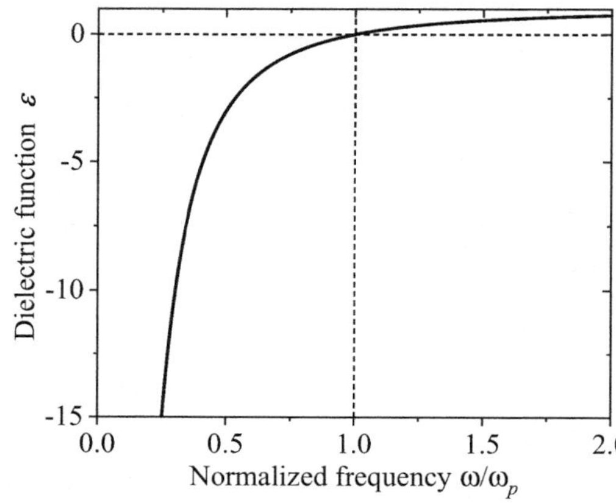

Fig. 2.6 Frequency spectrum of the dielectric function of a lossless metal predicted by the Drude model

frequency ω_p can be defined as the square root of $n_e^2/\varepsilon_0 m$. The dielectric function then reads

$$\varepsilon = 1 - \frac{\omega_p^2}{(\omega^2 + i\Gamma\omega)}. \tag{2.28}$$

The plasma frequency ω_p appears as a zero of the dielectric function when the medium is lossless (Fig. 2.6). The medium becomes fully reflective when $\omega < \omega_p$ since the dielectric function then takes negative values and its square root is a pure imaginary number. As we will soon show that the wave vector k appearing in the plane wave solution, e^{ikr}, is defined by the dispersion relation $k = \omega\sqrt{\epsilon}/c$, a negative dielectric function implies a pure imaginary wave vector and an evanescent wave. Consequently to the fact that the wave cannot propagate inside the medium, it is totally reflected.

2.2.2 Dispersion Relations for Polaritonic Materials

The electromagnetic dispersion relation can now be derived from the equation of propagation in a medium without current sources [10]:

$$\nabla^2 E - \mu_0\epsilon_0\mu\epsilon\frac{\partial^2 E}{\partial t^2} = 0. \tag{2.29}$$

As the material is assumed to have no magnetic response, the relative permeability $\mu = 1$. Note that by definition $\varepsilon_0\mu_0 = 1/c^2$, where μ_0 appears as the vacuum permeability. Considering the solution $\mathbf{E} = \mathbf{E}_0 \cdot \exp[i(kr - \omega t)]$ leads to the simple expression for the dispersion relation $k = \omega\sqrt{\epsilon}/c$. The dispersion relation of a polar material with a single resonance is reported in Fig. 2.7.

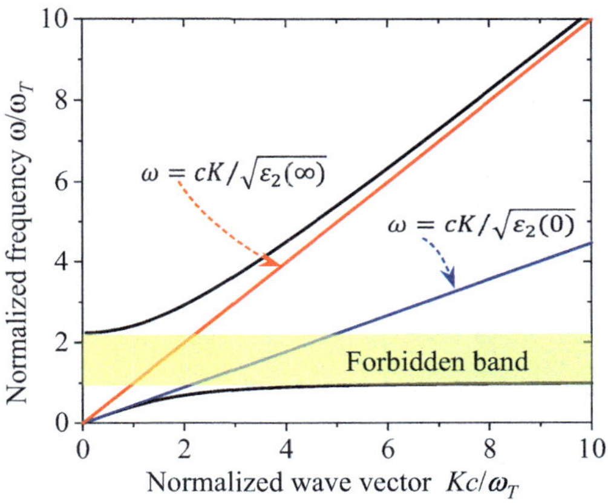

Fig. 2.7 Dispersion relation $k = \omega\sqrt{\epsilon}/c$ of a polar material with a single resonance. The forbidden band corresponds to the range where the dielectric function is negative, yielding imaginary solutions only. $\varepsilon(0)$ here corresponds to the static dielectric function taken at $\omega = 0$

The dispersion relation in Fig. 2.7 is typically the one of a mode hybridization or an anti-crossing, which combines the photon mode as the oblique branches and the horizontal optical phonon branches. This hybridization justifies the name of "Phonon-Polariton" for this mode. The intermediate region is marked by a band gap where the dielectric function is negative as evidenced in Fig. 2.6, where the same frequency units are used. In this specific range, only imaginary solutions can be found.

In the plasmon–polariton dispersion relation displayed in Fig. 2.8, a flat branch appears as seen in the phonon–polariton case in the low wave vector range. The number of modes and the density of states at the plasma frequency therefore peak, which explains the strong light absorption at this frequency. The real part of the dielectric function being negative at frequencies lower than ω_p, the wave vector, is a pure imaginary number and modes do not exist in this range.

2.2.3 *Surface Polariton Dispersions*

Having defined the dispersion of polaritons in a bulk material, we now turn to the description of the interaction between the surface and the polariton waves. As described in Fig. 2.9, we define the electric field inside and outside the material as E_1 and E_2, respectively, and set boundary conditions.

The wave vectors of the fields are decomposed into their common in-plane component K and their cross-plane components γ_i in the following form:

$$\mathbf{E_1} = \mathbf{E_1^0} e^{i(\mathbf{K}\cdot\mathbf{R}+\gamma_1 z)}, \tag{2.30}$$

$$\mathbf{E_2} = \mathbf{E_2^0} e^{i(\mathbf{K}\cdot\mathbf{R}-\gamma_2 z)}, \tag{2.31}$$

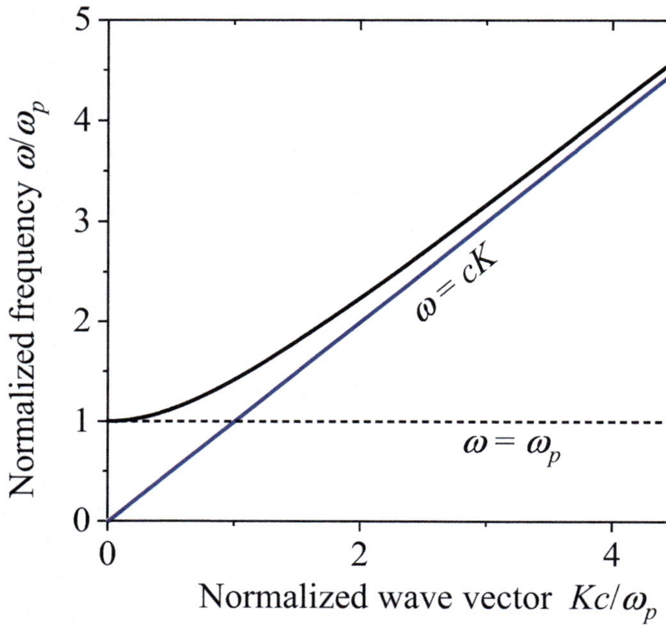

Fig. 2.8 Dispersion relation $k = \omega\sqrt{\epsilon}/c$ of the bulk plasmon–polariton corresponding to the dielectric function of Fig. 2.6. The dispersion relation converges to the light line at frequencies larger than $2\omega_p$. Below ω_p, the negative dielectric function yields pure imaginary wave vectors

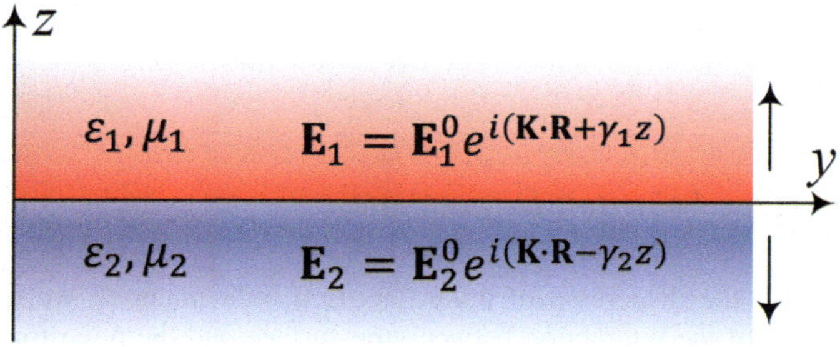

Fig. 2.9 Schematic of the interface between non-polaritonic and polaritonic materials with dielectric functions ε_1 and ε_2. The electric fields are defined by their amplitude and polarization s and p, E_i^0, and by their common in-plane wave vectors K and different out-of-plane wave vectors γ_i

where R denotes the in-plane position vector and the γ_i are the moduli of the out-of-plane wave vector. The s polarization parallel to the x axis is not solution if the permeabilities are equal. The fields with p polarization in the (y, z) plane:

$$\mathbf{E_1} = (E_{y,1}\boldsymbol{y} + E_{z,1}\boldsymbol{z})e^{i(\mathbf{K}\cdot\mathbf{R}+\gamma_1 z)}, \tag{2.32}$$

$$\mathbf{E_2} = (E_{y,2}\boldsymbol{y} + E_{z,2}\boldsymbol{z})e^{i(\mathbf{K}\cdot\mathbf{R}-\gamma_2 z)}, \tag{2.33}$$

have to comply with the continuity condition imposing equality of the tangential components of the fields at the interface: $E_{y,1} = E_{y,2}$. Combining this boundary condition with the Maxwell–Gauss equation $\nabla E = 0 = i\mathbf{k}.\mathbf{E_2} = i\mathbf{k}.\mathbf{E_1}$, which reduces to

$$KE_{y,2} - \gamma_2 E_{z,2} = KE_{y,1} + \gamma_1 E_{z,1}, \tag{2.34}$$

we obtain the following relation for the z components: $-\gamma_2 E_{z,2} = \gamma_1 E_{z,1}$. The continuity of the Electric Displacement at the interface $\epsilon_1 E_{z,1} = \epsilon_2 E_{z,2}$ then brings in the following relation on the perpendicular wave vectors and dielectric functions:

$$\epsilon_1\gamma_2 + \epsilon_2\gamma_1 = 0. \tag{2.35}$$

Taking into account that $\gamma_n = \varepsilon_n(\omega/c)^2 - K^2$, the solution of (2.35) for K is

$$K = \frac{\omega}{c}\sqrt{\frac{\varepsilon_1\varepsilon_2}{\varepsilon_1 + \epsilon_2}}. \tag{2.36}$$

Equation (2.36) is the dispersion of surface polaritons propagating along the interface of media 1 and 2. If medium 1 is air or vacuum, $\varepsilon_1 = 1$.

2.2.3.1 The Surface Phonon–Polariton

Using this latter equation, Fig. 2.10 reports the dispersion relation of surface phonon–polaritons based on the dielectric function of Fig. 2.6 with a single resonance. The photon-like branches at higher and lower frequencies actually have different slopes, the former being defined by the dielectric function equal to unity when ω tends to infinity. Those slopes clearly differ from the ones of the bulk phonon–polariton of Fig. 2.7. A more remarkable change compared to this latter dispersion also appears as the additional bands in the gap. This band is flat and therefore involves a large number of modes. Finally, even if alike, the surface phonon–polariton dispersion largely differs from that of the bulk modes.

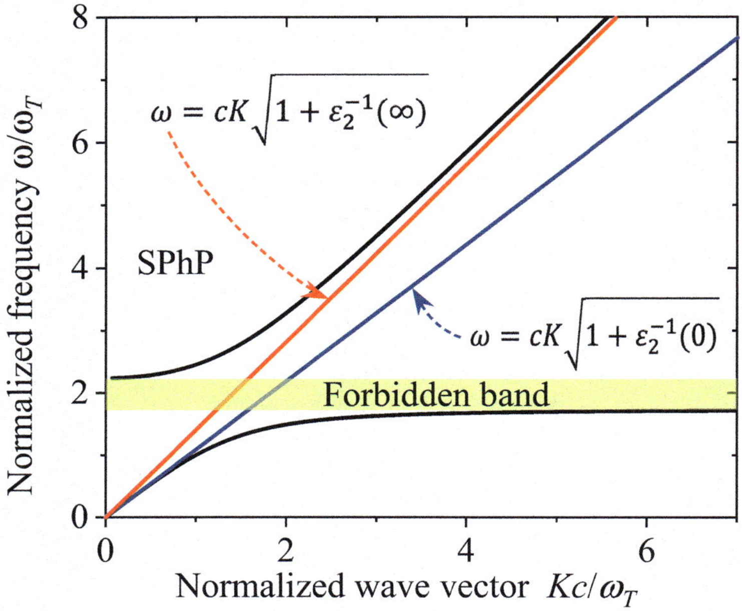

Fig. 2.10 Surface phonon–polariton dispersion relation derived from the dielectric function of Fig. 2.6

2.2.3.2 The Surface Plasmon–Polariton

The dispersion relation of surface plasmon–polaritons based on the dielectric function of Fig. 2.7 is provided in Fig. 2.11. The photon-like branches at higher frequencies follows a slope given by the product of the light velocity and the square root of two, as the dielectric function tends to unity. The lower frequency photon branch follows the light line. The gap where the wave vector is a pure imaginary number is defined in the interval $[\omega_p/\sqrt{2}, \omega_p]$. The $\omega_p/\sqrt{2}$ limit arises from the positivity condition $2 - \left(\omega_p/\omega\right)^2 > 0$ for the ratio below the square root in the dispersion relation $K = (\omega/c)\sqrt{\epsilon_2/(1+\epsilon_2)}$.

Both surface phonon and plasmon dispersions feature two branches separated by a gap. The contribution to heat transfer is mainly produced by the branch of lower frequencies. Indeed, the typical values of plasma frequencies ω_p for metals lie in the range between 100 and 1000 THz, which belongs to the optical and UV spectra, beyond the thermal InfraRed interval, 10 to 60 THz at room temperature. The plasmon resonance and the surface plasmon–polariton modes with larger frequencies therefore do not impact thermal radiation. Surface phonon–polaritons have resonances at some specific optical phonon frequencies, which correspond to the highest phonon frequencies of typically several tens of TeraHertz. The surface phonon–polariton resonance therefore belongs to the thermal InfraRed region, but

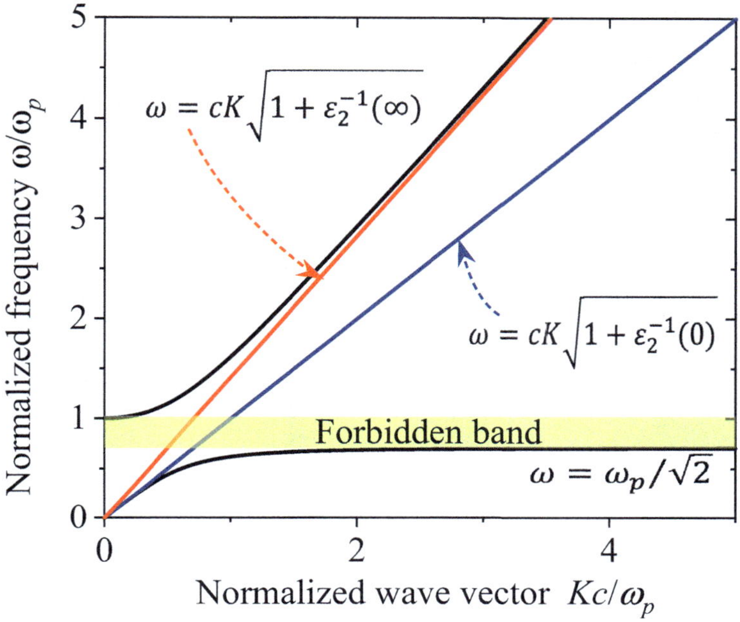

Fig. 2.11 Dispersion relation of the surface plasmon–polariton derived from the lossless dielectric function of Fig. 2.6

phonons do not exist beyond this frequency where electromagnetic modes are purely photons: They can be excited via electromagnetic sources such as lasers but not with thermal sources at reasonable temperatures. On another hand, we have already mentioned that the modes of the resonance, inside the Restrahlen band, will be very efficiently absorbed and actually have small propagation lengths, and their heat flux will not be predominant.

References

1. Planck, M.: Theory of Heat Radiation. New York (1914)
2. Volz, S., Lemonnier, D. Saulnier, J.-B., Clamped nanowire thermal conductivity based on phonon transport equation. Microscale Thermophys. Eng. **5**, 191–207 (2001)
3. Ziman, J.M.: Electrons and Phonons: The Theory of Transport Phenomena in Solids. Oxford University Press, Oxford (2001)
4. Zhang, Z.M.: Nano/Microscale Heat Transfer. Springer, Berlin (2020)
5. Howell, J.R., Pinar Menguc, M., Daun, K.: Thermal Radiation: An Introduction. CRC Press, Boca Raton (2023)
6. Tachikawa, S., Ordonez-Miranda, J., Jalabert, L., Wu, Y., Anufriev, R., Guo, Y., Kim, B., Fujita, H., Volz, S., Nomura, M.: Enhanced far-field thermal radiation through a polaritonic waveguide. Phys. Rev. Lett. **132**(18), 186904 (2024)

7. Chen, S.A., Narayanaswamy, A., Chen, G.: Surface phonon-polariton mediated thermal conductivity enhancement of amorphous thin films. Phys. Rev. B **72**, 155435 (2005)
8. Ordonez-Miranda, J., Tranchant, L., Kim, B., Chalopin, Y., Antoni, T., Volz, S.: Quantized thermal conductance of nanowires at room temperature due to zenneck surface-phonon polaritons, Phys. Rev. Lett. **112**, 055901 (2014)
9. Dresselhaus, M.S.: Solid State Physics, Part II, Optical Properties of Solids. Online
10. Jackson, J.D.: Classical Electrodynamics. Wiley, London (1998)

Chapter 3
Electromagnetic Field Equations

3.1 Maxwell's Equations

The Maxwell's equations describe the spatial and temporal distribution of the electromagnetic fields via the following four coupled equations:

$$\text{Coulomb's law} \qquad \nabla \cdot \mathbf{D} = \rho, \tag{3.1}$$

$$\text{Ampere's law} \qquad \nabla \times \mathbf{H} = \mathbf{J} + \frac{\partial \mathbf{D}}{\partial t}, \tag{3.2}$$

$$\text{Faraday's law} \qquad \nabla \times \mathbf{E} + \frac{\partial \mathbf{B}}{\partial t} = 0, \tag{3.3}$$

$$\text{Gauss's law} \qquad \nabla \cdot \mathbf{B} = 0, \tag{3.4}$$

where $\mathbf{E}$ (Vm^{-1}), $\mathbf{D}$ (Cm^{-2}), $\mathbf{H}$ (Am^{-1}), and $\mathbf{B}$ (T) are the electric, electric displacement, magnetic, and magnetic induction fields, respectively. The Maxwell's equations are written in units (mks) of the International System (SI), which is used throughout this book. The electric current density $\mathbf{J}$ (Am^{-2}) is related to the electric charge density ρ (Cm^{-3}) through the continuity equation

$$\nabla \cdot \mathbf{J} + \frac{\partial \rho}{\partial t} = 0, \tag{3.5}$$

which represents the conservation of local charge, as both $\mathbf{J}$ and ρ are generally functions of position $\mathbf{x}$ and time t. In words, the Coulomb's law in (3.1) establishes that the total electric flux passing through a closed surface is proportional to the total charge enclosed by that surface. Therefore, the Coulomb's law describes how electric charges create electric fields. The Gauss's law in (3.4) states that the total magnetic flux passing through a closed surface is zero. The Gauss's law

S. Volz, J. Ordonez-Miranda, *Heat Transport Driven by Surface Electromagnetic Waves*, Mechanical Engineering Series, https://doi.org/10.1007/978-3-031-66024-5_3

thus indicates that, unlike electric charges, there are no isolated magnetic charges (monopoles). The Ampere's law in (3.2) establishes that the circulation of the magnetic field around a closed loop is proportional to the total current passing through the loop plus and displacement current ($\partial \mathbf{D}/\partial t$). This latter term was added by Maxwell to ensure consistency with the conservation of charge in (3.5), for time-dependent fields. Finally, the Faraday's law in (3.3) describes how the time variation of the magnetic field induces an electric field and vice versa. The Maxwell's equations (3.1)–(3.4) thus describe the generation and propagation of the electric and magnetic fields, and they unify the electromagnetic phenomena into a coherent basis.

The continuity equation in (3.5) is contained in Maxwell's equations (3.1) and (3.2). Taking the divergence of (3.2), one obtains $\nabla \cdot (\nabla \times \mathbf{H}) = \nabla \cdot \mathbf{J} + \partial\left(\nabla \cdot \mathbf{D}\right)/\partial t = 0$, as the divergence of a curl is zero. The insertion of (3.1) into this relation yields (3.5), which shows the consistence of the fundamental charge conservation with Maxwell's equations. Conversely, by replacing $\nabla \cdot \mathbf{J} = -\partial\rho/\partial t$ into this latter relation, we obtain the Coulomb's law in (3.1). The divergence equation (3.1) is thus a direct consequence of the curl equation (3.2), provided that the charge conservation in (3.5) is satisfied. This is also the case of (3.4), which results directly from the divergence of Faraday's law in (3.3). Because of the dependence of (3.1) and (3.4) on (3.2) and (3.3), we are going to focus on these latter two curl equations to describe the behavior of the electromagnetic fields. For media supporting linear relationships between these fields, the solution of (3.2) and (3.3) can conveniently be found in the Fourier's space. Recalling the Fourier's transform of $\mathbf{F} = \mathbf{E}, \mathbf{D}, \mathbf{H}, \mathbf{B}$, or $\mathbf{J}$ is given by

$$\mathbf{F}(\mathbf{x}, t) = \int_{-\infty}^{\infty} \mathbf{F}(\mathbf{x}, \omega) e^{-i\omega t} d\omega. \tag{3.6}$$

We can rewrite the Maxwell's equations without the time derivatives, as follows:

$$\nabla \times \mathbf{H} + i\omega \mathbf{D} = \mathbf{J}, \tag{3.7}$$

$$\nabla \times \mathbf{E} - i\omega \mathbf{B} = 0, \tag{3.8}$$

where we have omitted the dependence of the fields on position $\mathbf{x}$ and frequency ω for the sake of simplicity. Any of these fields $\mathbf{F}(\mathbf{x}, \omega)$ is a complex vector function that satisfies the condition $\mathbf{F}(\mathbf{x}, -\omega) = \mathbf{F}^*(\mathbf{x}, \omega)$. This relation results from the reality ($\mathbf{F}(\mathbf{x}, t) = \mathbf{F}^*(\mathbf{x}, t)$) of the fields in time space, as established by (3.6). In addition, the sum of (3.6) with its complex conjugate yields $2\mathbf{F}(\mathbf{x}, t) = \int_{-\infty}^{\infty} \left[\mathbf{F}(\mathbf{x}, \omega) e^{-i\omega t} + \mathbf{F}^*(\mathbf{x}, \omega) e^{i\omega t}\right] d\omega = 2\int_{-\infty}^{\infty} \mathrm{Re}\left[\mathbf{F}(\mathbf{x}, \omega) e^{-i\omega t}\right] d\omega$, which justifies the utilization of the substitution $\mathbf{F}(\mathbf{x}, t) \to \mathrm{Re}\left[\mathbf{F}(\mathbf{x}, \omega) e^{-i\omega t}\right]$, provided that the integration over all frequencies is implicitly understood. Taking into account that the independent equations (3.7) and (3.8) involve four fields, for a given source current $\mathbf{J}(\mathbf{x}, \omega)$, two additional equations relating to these complex fields are required to determine them.

3.2 The Constitutive Relations

Considering that the electromagnetic fields are propagating in linear and isotropic media, the complex fields **D** and **B** are linear functions of **E** and **H**, as follows:

$$\mathbf{D}(\mathbf{x}, \omega) = \varepsilon(\omega)\mathbf{E}(\mathbf{x}, \omega), \tag{3.9}$$

$$\mathbf{B}(\mathbf{x}, \omega) = \mu(\omega)\mathbf{H}(\mathbf{x}, \omega), \tag{3.10}$$

where ε and μ are the electric permittivity and magnetic permeability of the medium supporting the fields, respectively. ε is a measure of the medium ability to support the formation of an electric field in response to an applied electric charge or voltage. A medium with higher electric permittivity thus supports a larger electric field. On the other hand, μ measures the medium ability to support the formation of a magnetic field in response to an applied magnetic flux. Media with higher magnetic permeability support a stronger magnetic field. These two fundamental properties ε and μ thus describe the macroscopic response of a medium to electric and magnetic fields, respectively. Given that ε and μ are, in general, complex functions of frequency, they are well-defined in the Fourier's space only, as established by (3.9) and (3.10). However, if these constitute parameters become independent of frequency, over a relevant interval of frequencies, they can also be defined in time space or for static fields. In any case, the reality condition $\mathbf{D}(\mathbf{x}, t) = \mathbf{D}^*(\mathbf{x}, t)$ applied to (3.6) along with (3.9) establishes that $\int_{-\infty}^{\infty} \varepsilon(\omega)\mathbf{E}(\mathbf{x}, \omega)e^{-i\omega t}d\omega = \int_{-\infty}^{\infty} \varepsilon^*(\omega)\mathbf{E}^*(\mathbf{x}, \omega)e^{i\omega t}d\omega$. By using the change of variable $\omega = -\omega'$ in the left-hand side of this relation and applying the identity $\mathbf{E}(\mathbf{x}, -\omega) = \mathbf{E}^*(\mathbf{x}, \omega)$, one obtains $\varepsilon(-\omega) = \varepsilon^*(\omega)$. Similarly, the reality of $\mathbf{B}(\mathbf{x}, t)$ implies that $\mu(-\omega) = \mu^*(\omega)$. This frequency dependence of the electric permittivity and magnetic permeability is known as their dispersive property.

The characterization of the electromagnetic properties of a medium by its permittivity and permeability is essential in macroscopic electrodynamics. For vacuum (free space), these properties are independent of frequency and take the following experimental values:

$$\varepsilon = \varepsilon_0 = 8.854\,187\,812 \times 10^{-12} \left(\mathrm{Fm}^{-1}\right), \tag{3.11}$$

$$\mu = \mu_0 = 1.256\,637\,062 \times 10^{-6} \left(\mathrm{NA}^{-2}\right). \tag{3.12}$$

These two constants define the speed of light in vacuum c, which is a universal constant given by

$$c = \frac{1}{\sqrt{\varepsilon_0 \mu_0}} = 299\,792\,458 \left(\mathrm{ms}^{-1}\right) \approx 3 \times 10^8 \left(\mathrm{ms}^{-1}\right). \tag{3.13}$$

From now on, we will limit ourselves to nonmagnetic media, for which (3.12) applies, and focus on the frequency dependence of their relative permittivity (also called dielectric function) $\varepsilon_r = \varepsilon/\varepsilon_0$. For the sake of simplicity, ε_r will simply be denoted by ε hereafter.

3.2.1 *Linear and Isotropic Medium*

Here we detail the models for the relative permittivity of dielectrics and metals exhibiting a linear and isotropic behavior.

3.2.1.1 Permittivity of Dielectric Materials

The frequency dependence of the relative permittivity of dielectric materials is described by the Lorentz model proposed in 1909 by the Dutch physicist Hendrik Lorentz [1]. This classical model considers that the dielectric electrons are bound to their respective atom nuclei via a hypothetical spring and its motion is damped by via a damper. The application of a uniform electric field $\mathbf{E(t)}$ pushes the electrons away from the nuclei and induces "non-symmetric electron clouds" forming electric dipoles with the positive nuclei, as shown in Fig. 3.1. Considering that the position $\mathbf{x}$ of a single electron with charge e and effective mass m can be described by Newton's second law, one obtains

$$\mathbf{F}_{\text{spring}} + \mathbf{F}_{\text{damping}} + \mathbf{F}_{\text{external}} = m\frac{d^2\mathbf{x}}{dt^2}, \tag{3.14}$$

where $\mathbf{F}_{\text{spring}} = -k\mathbf{x}$, $\mathbf{F}_{\text{damping}} = -m\Gamma d\mathbf{x}/dt$, and $\mathbf{F}_{\text{external}} = -e\mathbf{E}$ are the respective spring, damping, and external forces, with k and Γ being the spring

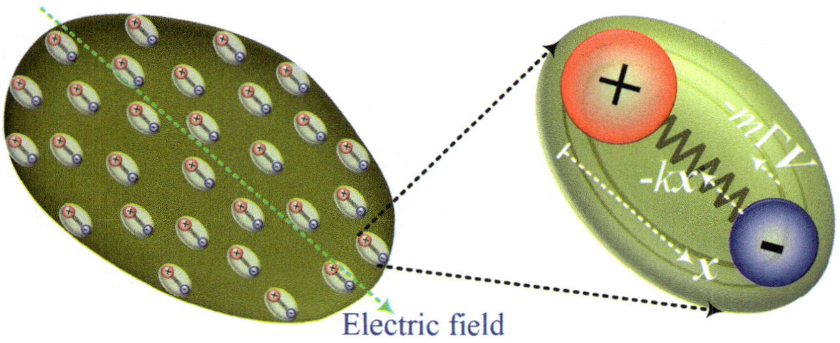

Fig. 3.1 Lorentz model electrons (blue points) bound to the atom nuclei (red points) through spring-like forces

constant and friction felt by the electron to move inside the material, respectively. After inserting these internal and external forces into (3.14), we find that the electron oscillations are driven by

$$\frac{d^2\mathbf{x}}{dt^2} + \Gamma\frac{d\mathbf{x}}{dt} + \omega_0^2\mathbf{x} = -\frac{e}{m}\mathbf{E}(t), \tag{3.15}$$

where $\omega_0 = \sqrt{k/m}$ is the electron resonance frequency. Taking the Fourier's transform defined in 3.6, in both sides of 3.15, one finds

$$\mathbf{x}(\omega) = \frac{e}{m}\frac{\mathbf{E}(\omega)}{\omega^2 - \omega_0^2 + i\Gamma\omega}. \tag{3.16}$$

The damping factor $\Gamma > 0$ thus ensures that the electron position is finite when the frequency of the external electric field matches the natural frequency of the electron's oscillations. The amplitude of these oscillations is however enhanced for $\omega = \omega_0$, which hence represents the material resonance frequency. The fact that (3.16) is a complex equation means there is a time delay (phase shift) between the driving electric field and the electron motion, due to the damping effect.

The electron displacement from the atom nuclei induces the formation of an electric dipole (see Fig. 3.1) with a moment $\mathbf{p} = -e\mathbf{x}$. Considering that the material contains n dipoles per unit volume and that each electron is acting with the same dipole moment, the polarization $\mathbf{P} = n\mathbf{p}$ takes the form

$$\mathbf{P}(\omega) = -\frac{\varepsilon_0\omega_p^2}{\omega^2 - \omega_0^2 + i\Gamma\omega}\mathbf{E}(\omega), \tag{3.17}$$

where the characteristic frequency $\omega_p = e\sqrt{n/m\varepsilon_0}$ represents the natural frequency at which the collection of electrons respond to an electric field. This material polarization represents the internal (induced) electrical field, whose superposition with the external one determines the electric displacement field $\mathbf{D}$ through the relation $\mathbf{D} = \varepsilon_0\varepsilon\mathbf{E} = \varepsilon_0\mathbf{E} + \mathbf{P}$. The combination of this relation with (3.17) yields the following relative permittivity:

$$\varepsilon(\omega) = 1 - \frac{\omega_p^2}{\omega^2 - \omega_0^2 + i\Gamma\omega}. \tag{3.18}$$

Note that the relative permittivity converges to $\varepsilon(0) = 1 + \left(\omega_p/\omega_0\right)^2$ and $\varepsilon(\infty) = 1$, for low and high enough frequencies, respectively. The independence of these real and positive values from the damping factor indicates that the complex nature of the permittivity appears in the neighborhood of the resonance ($\omega \sim \omega_0$) mainly. At high frequency ($\omega \to \infty$), materials usually exhibit a highly polarized environment (positive background of the atoms' nuclei) that is not well described by (3.18). This residual polarization can, however, be incorporated into the Lorentz model by considering $\varepsilon(\infty) \geq 1$ (usually $1 \leq \varepsilon(\infty) \leq 10$) as an adjustable parameter. In terms of $\varepsilon(0)$ and $\varepsilon(\infty)$, (3.18) can thus be written as follows:

$$\varepsilon(\omega) = \varepsilon(\infty) - \frac{[\varepsilon(0) - \varepsilon(\infty)]\,\omega_0^2}{\omega^2 - \omega_0^2 + i\Gamma\omega}. \tag{3.19}$$

The four parameters in (3.19) thus characterize the material response to the application of an electrical field. In particular, they can be related to the field propagation frequencies via the Maxwell equations. In the absence of free chargers ($\mathbf{J} = 0$) and in terms of the fields $\mathbf{E}$ and $\mathbf{B} = \mu_0\mathbf{H}$, (3.7), and (3.8) take the form

$$\nabla \times \mathbf{B} + \frac{i\omega}{c^2}\varepsilon\mathbf{E} = 0, \tag{3.20}$$

$$\nabla \times \mathbf{E} - i\omega\mathbf{B} = 0. \tag{3.21}$$

Combining (3.20) and (3.21) yields the following Helmholtz wave equation:

$$\left(\nabla^2 + \varepsilon k_0^2\right)\begin{bmatrix}\mathbf{E}\\ \mathbf{B}\end{bmatrix} = 0, \tag{3.22}$$

where $k_0 = \omega/c$. The wave equation (3.22) indicates that $\mathbf{E}$ and $\mathbf{B}$ propagate as plane waves and therefore they can be written as follows:

$$\mathbf{E} = \mathbf{E_0}e^{i\mathbf{k}\cdot\mathbf{x}}, \tag{3.23}$$

$$\mathbf{B} = \mathbf{B_0}e^{i\mathbf{k}\cdot\mathbf{x}}. \tag{3.24}$$

After inserting (3.23) and (3.24) into (3.22), one obtains the following dispersion relation for the wave vector $k = k(\omega)$

$$k = \sqrt{\varepsilon}k_0. \tag{3.25}$$

k_0 thus represents the wave vector in vacuum ($\varepsilon = 1$). Furthermore, the combination of (3.23) and (3.24) with (3.20) and (3.21) yields

$$\hat{k} \cdot \mathbf{D}_0 = 0, \tag{3.26}$$

$$\hat{k} \times \mathbf{E}_0 = c\mathbf{B}_0/\sqrt{\varepsilon}, \tag{3.27}$$

where the unit vector $\hat{k}$ and the electric displacement field $\mathbf{D}_0$ are defined by $\mathbf{k} = k\hat{k}$ and $\mathbf{D} = \mathbf{D}_0e^{i\mathbf{k}\cdot\mathbf{x}}$, respectively. Taking into account that the fields $\mathbf{E}$, $\mathbf{D}$, and $\mathbf{P}$ are parallel, as the relative permittivity is a scalar (i.e., ε is not a tensor), (3.26) and (3.27) establish that the propagation of these fields is driven by two modes, mutually orthogonal for $\varepsilon \to \infty$. These modes are defined by

$$\mathbf{D} = 0 \qquad \text{or} \qquad \mathbf{P} \parallel \mathbf{k}, \tag{3.28}$$

$$\mathbf{E} = 0 \qquad \text{or} \qquad \mathbf{P} \perp \mathbf{k}. \tag{3.29}$$

A longitudinal mode can therefore be defined by a nonzero polarizability $\mathbf{P} \parallel \mathbf{k}$, for which (3.28) predicts that $\mathbf{D} = 0$. This zero electric displacement yields

$$\varepsilon = 0 \qquad \text{and} \qquad \mathbf{P} = -\varepsilon_0 \mathbf{E}. \tag{3.30}$$

On the other hand, according to (3.29), the transverse mode is defined by a nonzero polarizability $\mathbf{P} \perp \mathbf{k}$ and an electric field $\mathbf{E} = 0$. As this latter condition in (3.29) was derived for a very large permittivity, the transverse mode is characterized by

$$\varepsilon = \infty \qquad \text{and} \qquad \mathbf{P} = \mathbf{D}. \tag{3.31}$$

The longitudinal and transverse modes are thus defined by a permittivity $\varepsilon = 0$ and $\varepsilon = \infty$, respectively. Both modes have an associated net polarization, so they are expected to produce long-range electromagnetic fields able to propagate over distances much longer than the inter-atom ones. Assuming that these permittivity values appear at the frequency $\omega = \omega_L$, for the longitudinal mode, and $\omega = \omega_T$, for the transverse mode, the Lorentz model in (3.19) predicts that their values are given by

$$\omega_L = \sqrt{\frac{\varepsilon(0)}{\varepsilon(\infty)}\omega_0^2 - \left(\frac{\Gamma}{2}\right)^2} - \frac{i\Gamma}{2} \quad \rightarrow \quad \omega_0\sqrt{\frac{\varepsilon(0)}{\varepsilon(\infty)}}, \tag{3.32}$$

$$\omega_T = \sqrt{\omega_0^2 - \left(\frac{\Gamma}{2}\right)^2} - \frac{i\Gamma}{2} \quad \rightarrow \quad \omega_0. \tag{3.33}$$

The frequencies of the longitudinal and transverse modes are thus complex, with an imaginary part equal to the half of the damping factor $\Gamma > 0$. For an ideal lossless material ($\Gamma = 0$), however, they become real and satisfy the Lyddane–Sachs–Teller (LST) relation $\omega_L^2/\omega_T^2 = \varepsilon(0)/\varepsilon(\infty)$. For a lossy material ($\Gamma > 0$), on the other hand, (3.32) and (3.33) establish that $|\omega_L|^2 = [\varepsilon(0)/\varepsilon(\infty)]\,\omega_0^2$ and $|\omega_T|^2 = \omega_0^2$, where the vertical bars stand for the frequency amplitude. For realistic materials, the LST relation can hence be generalized, as follows:

$$\frac{|\omega_L|^2}{|\omega_T|^2} = \frac{\varepsilon(0)}{\varepsilon(\infty)}. \tag{3.34}$$

For polar materials, such as Boron Nitride (BN) and Silicon Carbide (SiC), the low-frequency permittivity is generally larger than the high-frequency one ($\varepsilon(0) > \varepsilon(\infty)$), and therefore $|\omega_L| > |\omega_T|$. On the other hand, for non-polar materials, such as silicon and germanium, the amplitudes of the longitudinal and transverse

Table 3.1 Lorentz model parameters of polar materials*

Dielectric	$\omega_L/2\pi$ (THz)	$\omega_T/2\pi$ (THz)	$\Gamma/2\pi$ (THz)	ε_∞
Boron Nitride (BN)	38.83	31.35	0.16	4.46
Silicon Carbide (SiC)	28.97	23.71	0.14	6.7
Silicon Dioxide (SiO_2)				
First resonance	14.99	14.24	0.19	2.0
Second resonance	37.24	32.95	1.42	2.0
Hexagonal Boron Nitride (hBN)				
Ordinary axis	48.22	41.06	0.15	4.87
Extraordinary axis	24.83	23.40	0.12	2.95

* These estimated values may vary depending on various factors, such as temperature and impurities in the dielectric material

frequencies degenerate ($|\omega_L| = |\omega_T|$), and thus $\varepsilon(0) = \varepsilon(\infty)$. In this latter case, the material permittivity becomes independent of frequency and is equal to the constant $\varepsilon = \varepsilon(\infty)$, as established by the Lorentz model in (3.19). For any case, the generalized LST relation in (3.34) allows rewriting this model in the following usual form:

$$\varepsilon(\omega) = \varepsilon_\infty \left(1 - \frac{\omega_L^2 - \omega_T^2}{\omega^2 - \omega_T^2 + i\Gamma\omega}\right), \tag{3.35}$$

where we have used the notation $\varepsilon_\infty = \varepsilon(\infty)$, $|\omega_L| \rightarrow \omega_L$, and $|\omega_T| \rightarrow \omega_T$ for the sake of brevity. The Lorentz model in (3.35) provides a useful framework for understanding the electric response of linear and isotropic dielectrics. Even though this simplified model does not capture all the complexities of real-world dielectrics, it is the basis for more advanced theories and models. The experimental values of the four parameters: ε_∞, ω_L, ω_T, and Γ are summarized in Table 3.1, for some typical polar dielectrics.

According to (3.35), the frequency dependence of the real (ε_R) and imaginary (ε_I) parts of the dielectric permittivity $\varepsilon = \varepsilon_R + i\varepsilon_I$ is given by

$$\frac{\varepsilon_R}{\varepsilon_\infty} = 1 - \frac{\left(\omega_L^2 - \omega_T^2\right)\left(\omega^2 - \omega_T^2\right)}{\left(\omega^2 - \omega_T^2\right)^2 + (\Gamma\omega)^2}, \tag{3.36}$$

$$\frac{\varepsilon_I}{\varepsilon_\infty} = \frac{\left(\omega_L^2 - \omega_T^2\right)\Gamma\omega}{\left(\omega^2 - \omega_T^2\right)^2 + (\Gamma\omega)^2}. \tag{3.37}$$

Figure 3.2a shows the typical behavior of ε_R and ε_I as functions of frequency. Note that the energy loss ($\varepsilon_I > 0$) is high for frequencies near the resonances (peak and dip) of ε_R and vanishes for other frequencies. Far high and low frequencies from these resonances, the permittivity can hence be replaced by its real, positive, and constant values ε_∞ and $\varepsilon_\infty(\omega_L/\omega_T)^2$, respectively. The maximum and minimum

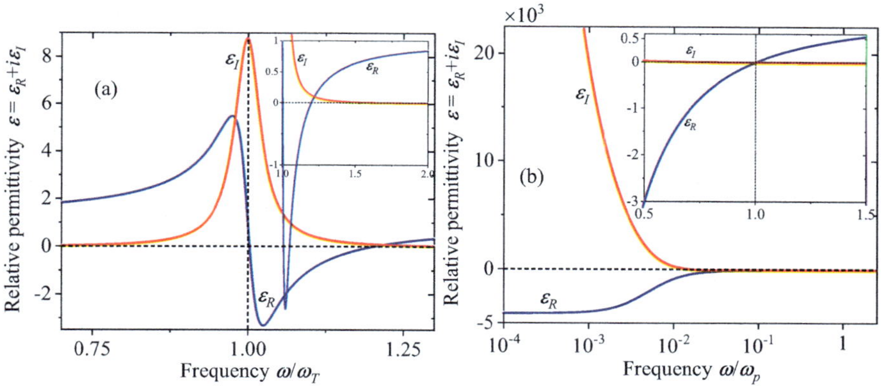

Fig. 3.2 Relative permittivity of (**a**) a metal and (**b**) polar dielectric predicted by the Drude and Lorentz models, respectively. Calculations were done for the typical values $\Gamma = 5 \times 10^{-3}\omega_p$ in (**a**) and $\omega_L = 1.2\omega_T$, $\Gamma = 5 \times 10^{-2}\omega_T$, and $\varepsilon_\infty = 1$ in (**b**)

of ε_R appear at the respective frequencies $\omega_+ = \omega_T\sqrt{1 - \Gamma/\omega_T}$ and $\omega_- = \omega_T\sqrt{1 + \Gamma/\omega_T}$ and take the following values:

$$\frac{\varepsilon_{R,\max}}{\varepsilon_\infty} = 1 + \frac{\omega_L^2 - \omega_T^2}{\Gamma(2\omega_T - \Gamma)}, \tag{3.38}$$

$$\frac{\varepsilon_{R,\min}}{\varepsilon_\infty} = 1 - \frac{\omega_L^2 - \omega_T^2}{\Gamma(2\omega_T + \Gamma)}. \tag{3.39}$$

Equation (3.39) establishes that ε_R takes negative values because $\omega_L > \omega_T$, such that the amplitude of its dip increases with the square of ω_L. This behavior is also exhibited by the sum and difference of the maximum and minimum values of ε_R, which are given by

$$\frac{\varepsilon_{R,\max} + \varepsilon_{R,\min}}{2\varepsilon_\infty} = \frac{3\omega_T^2 + \omega_L^2}{4\omega_T^2 - \Gamma^2}, \tag{3.40}$$

$$\frac{\varepsilon_{R,\max} - \varepsilon_{R,\min}}{\varepsilon_\infty} = \frac{4\omega_T}{\Gamma}\frac{\omega_L^2 - \omega_T^2}{4\omega_T^2 - \Gamma^2}. \tag{3.41}$$

The permittivity difference $\varepsilon_{R,\max} - \varepsilon_{R,\min}$ is thus driven by the frequency ω_T and takes higher values for smaller losses (damping factor Γ). On the other hand, the two zeros of ε_R shown in Fig. 3.2a show up at the frequencies

$$\Omega_+ = \omega_L\left(1 - \frac{0.5\Gamma^2}{\omega_L^2 - \omega_T^2}\right), \tag{3.42}$$

$$\Omega_- = \omega_T \left(1 + \frac{0.5\Gamma^2}{\omega_L^2 - \omega_T^2}\right), \tag{3.43}$$

which are valid for an approximation up to $\Gamma/\omega_T)^2 \ll 1$, as is usually the case of a wide variety of materials. The real part of the permittivity takes negative values for all frequencies within the interval $\Omega_- < \omega < \Omega_+$, whose boundaries, according to (3.42) and (3.43), fulfill the following constraints:

$$\frac{\Omega_+}{\omega_L} + \frac{\Omega_-}{\omega_T} = 2, \tag{3.44}$$

$$\Omega_+ - \Omega_- = \omega_L - \omega_T - \frac{0.5\Gamma^2}{\omega_L - \omega_T}. \tag{3.45}$$

The frequency window in which $\varepsilon_R < 0$ is thus nearly equal to the frequency difference $\omega_L - \omega_T$, for materials with low absorption ($\Gamma \ll \omega_T$). For a linear approximation on Γ/ω_T, $\Omega_+ = \omega_L$, $\Omega_- = \omega_T$, and the area A_R between the curves $\varepsilon_R < 0$ and $\varepsilon_R = 0$ can be expressed in a concise way, as follows:

$$A_R = \int_{\Omega_-}^{\Omega_+} \varepsilon_R(\omega) d\omega \approx \varepsilon_\infty(\omega_L - \omega_T)\left[1 - \frac{(\omega_L + \omega_T)}{\omega_T} \ln\left(\frac{4\omega_T}{\Gamma}\frac{\omega_L - \omega_T}{\omega_L + \omega_T}\right)\right]. \tag{3.46}$$

The area A_R is thus explicitly driven by the frequency difference $\omega_L - \omega_T$, and its negative values increase as the energy loss decreases. The negativity of the real part of the permittivity increases logarithmically with the ratio ω_T/Γ, as shown in (3.46). On the other hand, the positivity of the imaginary part of the permittivity can be evaluated via the area A_I between the curves $\varepsilon_I > 0$ and $\varepsilon_I = 0$. According to (3.37) and Fig. 3.2a, this area is given by

$$A_I = \int_0^\infty \varepsilon_I(\omega) d\omega = \varepsilon_\infty \frac{\omega_L^2 - \omega_T^2}{\sqrt{4\omega_T^2 - \Gamma^2}}\left[\frac{\pi}{2} + \arctan\left(\frac{2\omega_T^2 - \Gamma^2}{\Gamma\sqrt{4\omega_T^2 - \Gamma^2}}\right)\right], \tag{3.47}$$

$$A_I = \varepsilon_\infty \frac{\omega_L^2 - \omega_T^2}{2\omega_T}\left[\pi - \frac{\Gamma}{\omega_T} + \frac{5\pi}{16}\left(\frac{\Gamma}{\omega_T}\right)^2 + \cdots\right], \tag{3.48}$$

where the Taylor series expansion in (3.48) holds for $\Gamma \ll \omega_T$. In contrast to A_R, A_I is driven by the difference $\omega_L^2 - \omega_T^2$ of the square of the longitudinal and transverse frequencies. Note that A_I becomes independent of the energy losses, as $\Gamma/\omega_T \to 0$. This behavior is reasonable, given that the reduction of ε_I, for $\Gamma/\omega_T \to 0$, is counterbalanced by its divergence at $\omega = \omega_T$, as established by (3.37). For a finite

$\Gamma > 0$ and an approximation up to $(\Gamma/\omega_T)^2 \ll 1$, the maximum of ε_I appears at the frequency $\omega = \omega_T \left[1 - 0.5(\Gamma/2\omega_T)^2\right]$ and takes the following value:

$$\varepsilon_{I,\max} = \varepsilon_\infty \frac{\omega_L^2 - \omega_T^2}{\Gamma\omega_T} \left[1 + \left(\frac{\Gamma}{4\omega_T}\right)^2\right]. \tag{3.49}$$

The peak of the imaginary part of the permittivity thus increases as the energy loss decreases due to its occurrence close to the resonance frequency ω_T. Note that, for a linear approximation on Γ/ω_T, (3.41) and (3.49) establish that the extreme values of the real and imaginary parts of the permittivity are related by

$$\varepsilon_{I,\max} \approx \varepsilon_{R,\max} - \varepsilon_{R,\min} \approx \varepsilon_\infty \frac{\omega_L^2 - \omega_T^2}{\Gamma\omega_T}, \tag{3.50}$$

which indicates that the amplitude of the ε_I peak is approximately equal to the distance between the peak and dip of ε_R.

3.2.1.2 Permittivity of Metals

For metals, the frequency dependence of the relative permittivity can be described by the Drude model proposed in 1900 by the German physicist Paul Drude [2]. This simple classical model treats a metal exposed to an external electric field like a sea of free electrons bouncing and re-bouncing off heavier and relatively immobile positive ions, as shown in Fig. 3.3. The collisions of electrons with these ions cause their scattering and loss of momentum, which are responsible for the metal resistivity. Considering that the wavelength of the external electric field is much longer than the mean free path between collisions, the free electrons "feel" a homogeneous field. As the free electrons are not anchored to their nuclei (no springs), the spring force in (3.14) reduces to zero. Under this condition, the natural frequency $\omega_0 = \sqrt{k/m} = 0$, and the Lorentz model in (3.18) reduces to the Drude model

$$\varepsilon = 1 - \frac{\omega_p^2}{\omega^2 + i\Gamma\omega}. \tag{3.51}$$

For metals, the frequency $\omega_p = e\sqrt{n/m\varepsilon_0}$ is called plasma frequency because of its analogy with the oscillation frequency of the charged particles of a plasma, a state of matter consisting of an ionized gas containing free electrons and positive ions. As the concentration of electrons increases or their effective mass decreases, ω_p becomes higher, indicating a faster collective response. The plasma frequency thus measures the speed at which the information can propagate through a metal.

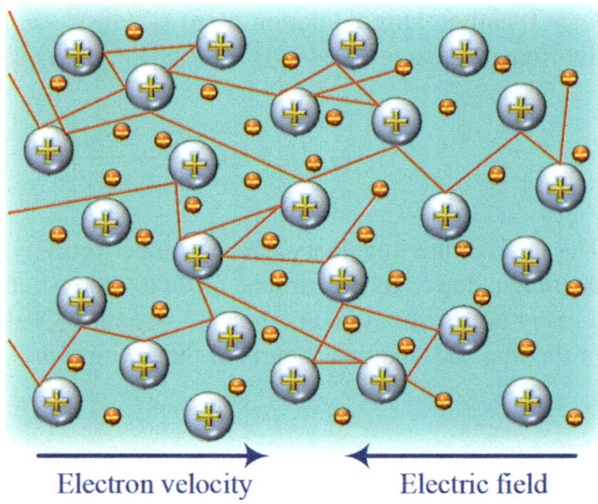

Fig. 3.3 Drude model electrons (red points) bouncing off heavier and nearly stationary positive ions (gray points). The red lines stand for the electron paths

The real (ε_R) and imaginary (ε_I) parts of $\varepsilon = \varepsilon_R + i\varepsilon_I$ in (3.51) are given by

$$\varepsilon_R = 1 - \frac{\omega_p^2}{\omega^2 + \Gamma^2}, \tag{3.52}$$

$$\varepsilon_I = \frac{\Gamma\omega_p^2}{\omega\left(\omega^2 + \Gamma^2\right)}. \tag{3.53}$$

The values of $\varepsilon_I = (1 - \varepsilon_R)\Gamma/\omega$ are thus driven by Γ, which accounts for the dissipative nature (energy loss) due to electron scattering in metals. The positive imaginary part ($\varepsilon_I > 0$) thus represents the dissipation of energy from the electromagnetic wave into the medium.[1] For high frequencies ($\Gamma \ll \omega < \omega_p$), the energy loss becomes negligible ($\varepsilon_I \to 0$), and the permittivity reduces to $\varepsilon \approx \varepsilon_R = 1 - (\omega_p/\omega)^2$. The negative $\varepsilon_R < 0$ for high frequency $\omega < \omega_p$ is essential to keep the metal ability to support the propagation of polaritons. The values of ε_R become more negative for lower frequencies until reaching its saturation $\varepsilon_R \to 1 - (\omega_p/\Gamma)^2$, for low enough frequencies ($\omega \ll \Gamma$), as shown in Fig. 3.2b. In this latter frequency regime, $\varepsilon_I \propto 1/\omega$, and hence the electron energy loss increases without limit. This physical drawback of the Drude model indicates that its simplified representation of the electrons' behavior does not capture all the complexities of real metals. It assumes that the electrons are noninteracting and neglect quantum mechanical effects. However, it provides a useful starting point for understanding the electromagnetic response of metals and has been influential in the development of more advanced models. The estimated values of the Drude model parameters for same metals are summarized in Table 3.2.

[1] If $\varepsilon_I < 0$, the medium gives energy to the wave and amplification occurs, as in a laser or maser [3].

Table 3.2 Plasma frequency and damping factor of typical metals*

Metal	$\omega_p/2\pi$ (THz)	$\Gamma/2\pi$ (THz)
Aluminum	2753	10–592
Copper	2292	16–318
Silver	2196	16–318
Gold	2150	16–318
Iron	1703	16–160
Titanium	1579	16–160

* These approximate values may vary depending on various factors, such as temperature and impurities in the metal

Taking into account that the electric current density is related to the polarizability via the relation $\mathbf{J} = -ned\mathbf{x}/dt = d\mathbf{P}/dt$, the relative permittivity in (3.51) can be linked to the electrical conductivity σ of metals. In the Fourier frequency space, this relation takes the form $\mathbf{J} = -i\omega\mathbf{P}$, which in combination with (3.17) and the metal condition $\omega_0 = 0$ yields the Ohm's law

$$\mathbf{J}(\mathbf{x}, \omega) = \sigma(\omega)\mathbf{E}(\mathbf{x}, \omega), \tag{3.54}$$

$$\sigma(\omega) = \frac{\varepsilon_0\omega_p^2}{\Gamma - i\omega}, \tag{3.55}$$

which is a constitutive relationship between the internal current density and the external electric field. Equation (3.55) is the classical Drude model for the AC electrical conductivity of metals [2]. As the typical values of Γ are greater than tens of terahertz (see Table 3.2), (3.55) indicates that the electric conductivity of metals essentially becomes real and positive, for frequencies ($\omega \ll \Gamma$) lower than 1 THz. By contrast, for high frequencies ($\omega \gg \omega_p > \Gamma$), the metal behaves like a lossy dielectric, with a vanishing electrical conductivity. By inserting the polarizability $\mathbf{P} = -\mathbf{J}/i\omega = -\sigma\mathbf{E}/i\omega$ into the relation $\mathbf{D} = \varepsilon_0\varepsilon\mathbf{E} = \varepsilon_0\mathbf{E} + \mathbf{P}$, we obtain the fundamental relationship between the relative permittivity and the electric conductivity

$$\varepsilon = 1 + \frac{i\sigma}{\omega\varepsilon_0}. \tag{3.56}$$

Interestingly, the same result can be obtained from (3.7) $\nabla \times \mathbf{H} = \mathbf{J} - i\omega\mathbf{D} = (\sigma - i\omega\varepsilon_0)\,\mathbf{E} = -i\omega\varepsilon_0\varepsilon\mathbf{E}$. This fact indicates that we can disregard the effect of the electrical conductivity, provided that we define the relative permittivity by (3.56). As the combination of (3.55) and (3.56) reduces to the Drude model for the permittivity in (3.51), we conclude that this permittivity model contains the effect of the electric conductivity.

3.2.2 Dispersion and Energy Absorption

According to (3.23) and (3.25), the propagation and attenuation of the electric field wave in a medium of relative permittivity $\varepsilon = \varepsilon_R + i\varepsilon_I$ is driven by the complex wave vector k given by

$$\frac{k}{k_0} = \sqrt{\varepsilon} = \sqrt{\varepsilon_R + i\varepsilon_I} = \frac{\beta_+ + i\beta_-}{\sqrt{2}}, \tag{3.57}$$

where the parameters $\beta_\pm$ are determined by squaring (3.57) and equating the real and imaginary parts of the obtained equation. Taking into account that $\varepsilon_I > 0$, this procedure yields

$$\beta_\pm = \sqrt{|\varepsilon| \pm \varepsilon_R}. \tag{3.58}$$

The imaginary part of k is hence driven by the imaginary part ε, such that its value vanishes in absence of absorption ($\varepsilon_I = 0$). This behavior can also be observed in terms of the amplitude $|k| = k_0\sqrt{\varepsilon_R^2 + \varepsilon_I^2} = k_0\sqrt{|\varepsilon|}$ and phase $\theta = \arctan(\beta_-/\beta_+)$ of $k = |k|e^{i\theta}$. Taking the tangent of 2θ, we obtain $\tan(2\theta) = \tan[2\arctan(\beta_-/\beta_+)] = 2(\beta_-/\beta_+)/\left[1 - (\beta_-/\beta_+)^2\right] = 2\beta_-\beta_+/\left(\beta_+^2 - \beta_-^2\right) = \varepsilon_I/\varepsilon_R$, which explicitly shows that the phase $\theta = 0.5\arctan(\varepsilon_I/\varepsilon_R)$ becomes zero for $\varepsilon_I = 0$. For an electric field propagating along the x axis, as shown in Fig. 3.4, (3.23) takes the form

$$\mathbf{E} = \mathbf{E}_0 e^{ikx} = \mathbf{E}_0 e^{k_0(i\beta_+ - \beta_-)x/\sqrt{2}}, \tag{3.59}$$

which explicitly shows that the electric field propagates as an attenuated wave with wavelength $\lambda = 2\sqrt{2}\pi/(k_0\beta_+)$ and attenuation constant driven by $\beta_- > 0$. By defining the intensity $I(x) = |\mathbf{E}(x)|^2$ of the electric field as the square of its amplitude, (3.59) yields

$$I(x) = I(0)e^{-\alpha x}, \tag{3.60}$$

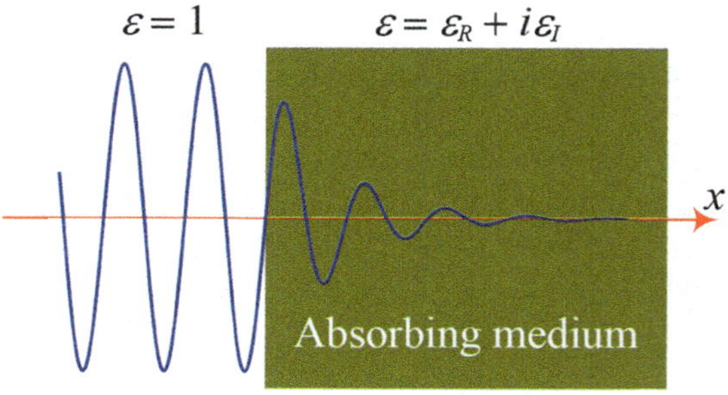

Fig. 3.4 Decaying waves of the electric field propagating inside an absorbing medium ($\mathrm{Im}(\varepsilon) > 0$)

where $\alpha = \sqrt{2}k_0\beta_-$ is known as the absorption coefficient. The electric field intensity thus falls off exponentially, and its value reduces by a factor of $e^{-1} \approx 0.368$ after traveling a distance α^{-1}. According to (3.58), the absorption coefficient is determined by the material permittivity via the following relation:

$$\alpha = \sqrt{2}k_0\sqrt{|\varepsilon| - \varepsilon_R}. \tag{3.61}$$

For materials with low absorption ($\varepsilon_I \ll |\varepsilon_R|$) and an approximation up to $(\varepsilon_I/\varepsilon_R)^2$, the permittivity magnitude $|\varepsilon| = \sqrt{\varepsilon_R^2 + \varepsilon_I^2} \approx |\varepsilon_R|\left[1 + 0.5\,(\varepsilon_I/\varepsilon_R)^2\right]$. Therefore, for frequencies in which $\varepsilon_R > 0$, the absorption coefficient in (3.61) reduces to

$$\alpha \approx k_0 \frac{\varepsilon_I}{\sqrt{\varepsilon_R}}, \tag{3.62}$$

which reveals the direct relation $\alpha \sim \varepsilon_I \sim \Gamma$, as established by the Drude and Lorentz models in (3.51) and (3.35), respectively. The damping factor Γ thus drives the energy absorption of the field wave by the material supporting its propagation. By contrast, for $\varepsilon_R < 0$, the leading term in (3.61) becomes independent of ε_I and takes the form

$$\alpha \approx 2k_0\sqrt{|\varepsilon_R|}. \tag{3.63}$$

The energy absorption of materials is thus mainly determined by the real part of their permittivity, for frequencies in which $\varepsilon_R < 0$ and $\varepsilon_I \ll |\varepsilon_R|$. On the other hand, for frequencies of high absorption ($\varepsilon_I \gg \varepsilon_R$) and a linear approximation on $\varepsilon_R/\varepsilon_I$, (3.61) simplifies to

$$\alpha \approx k_0\sqrt{2\varepsilon_I}\left(1 - \frac{\varepsilon_R}{2\varepsilon_I}\right), \tag{3.64}$$

which again reveals that the energy absorption is driven by the imaginary part of the permittivity and enhanced when real part of the permittivity is negative. Materials with higher $\varepsilon_I > 0$ and lower $\varepsilon_R < 0$ thus exhibit the stronger energy absorption. The frequency spectrum of the absorption coefficient predicted by (3.61) for dielectrics and metals following the respective Lorentz and Drude models is shown in Fig. 3.5. While the absorption of dielectrics appears around the neighborhood of the resonance frequency $\omega = \omega_T$ mainly, the one of metals pretty much vanishes for $\omega > \omega_p$. For $\omega < \omega_p$, the absorption coefficient of metals increases monotonically as the frequency decreases, due to the increasing imaginary part of their permittivity and the negative values of its real part, as shown in Fig. 3.2.

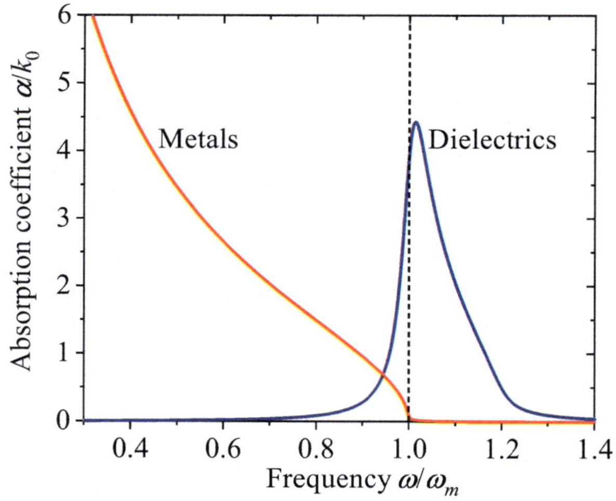

Fig. 3.5 Normalized absorption coefficient of metals and dielectrics predicted by the Drude ($\omega_m = \omega_p$) and Lorentz ($\omega_m = \omega_T$) models, respectively. Calculations were done with (3.61) and the parameters indicated in the caption of Fig. 3.2

3.2.3 Anisotropic Medium

A medium is anisotropic when the polarization of its atoms and molecules depends on the direction of the electric field that we apply to induce it. For this type of media, the relative permittivity is no longer a scalar, but rather a tensor $\bar{\bar{\varepsilon}}$ (matrix) relating the electric displacement and electric field as follows:

$$\mathbf{D}(\mathbf{x},\omega) = \varepsilon_0 \bar{\bar{\varepsilon}}(\omega)\mathbf{E}(\mathbf{x},\omega), \qquad \bar{\bar{\varepsilon}} = \begin{bmatrix} \varepsilon_{11} & \varepsilon_{12} & \varepsilon_{13} \\ \varepsilon_{21} & \varepsilon_{22} & \varepsilon_{23} \\ \varepsilon_{31} & \varepsilon_{32} & \varepsilon_{33} \end{bmatrix}, \tag{3.65}$$

where $\varepsilon_{ij} = \varepsilon_{ji}$ are the elements of the permittivity symmetric matrix describing the anisotropic behavior of the medium in different directions. This symmetry condition arises from the energy conservation and allows to diagonalize $\bar{\bar{\varepsilon}}$ by choosing a suitable orientation of the coordinate axes. These axes xyz in which the permittivity tensor is a diagonal matrix are called principal axes of the material and enable describing the fields in (3.65) via the following matrix equation:

$$\begin{bmatrix} D_x \\ D_y \\ D_z \end{bmatrix} = \varepsilon_0 \begin{bmatrix} \varepsilon_{xx} & 0 & 0 \\ 0 & \varepsilon_{yy} & 0 \\ 0 & 0 & \varepsilon_{zz} \end{bmatrix} \begin{bmatrix} E_x \\ E_y \\ E_z \end{bmatrix}, \tag{3.66}$$

which establishes that $D_n = \varepsilon_0 \varepsilon_{nn} E_n$, for the fields' component along the axis n. If the diagonal elements of the permittivity are all different ($\varepsilon_{xx} \neq \varepsilon_{yy} \neq \varepsilon_{zz}$), the material is called biaxial medium, as is the case of the van der Waals semiconductor α-phase molybdenum trioxide ($\alpha-MoO_3$). On the other hand, if two of the permittivity components are equal and one is different, the material is named uniaxial medium. If, for instance, $\varepsilon_{xx} \neq \varepsilon_{yy} = \varepsilon_{zz}$, the x axis is called extraordinary axis, while the y and z axes are called ordinary axes. Many materials such as hexagonal boron nitride (hBN), quartz, mica, and calcite are uniaxial, and the elements of their diagonal permittivity are well described by the Lorentz model in (3.35). Table 3.1 summarizes the values of the Lorentz parameters for the ordinary and extraordinary components of the hBN relative permittivity.

3.3 Boundary Conditions

The electromagnetic fields are uniquely determined by the solution of the Maxwell's equations, the associated constitutive relations (3.9) and (3.10), and the boundary conditions. These conditions can conveniently be derived by transforming the Maxwell's equations from their differential form in (3.1)–(3.4) into their integral one. This transformation is made possible by the following divergence and Stokes' theorems:

$$\int_V \nabla \cdot \mathbf{F} dV = \int_S \mathbf{F} \cdot \hat{\mathbf{n}} dS, \tag{3.67}$$

$$\int_A (\nabla \times \mathbf{F}) \cdot \hat{\mathbf{t}} dA = \int_C \mathbf{F} \cdot d\mathbf{l}, \tag{3.68}$$

where $\mathbf{F}$ is a well-behaved vector function, V is a volume in space, S is the closed surface bounding V, and $\hat{\mathbf{n}}$ is a unit vector normal to the surface element dS pointing outward from the enclosed volume, as shown in Fig. 3.6. Similarly, A is an open surface, C is a contour bounding it with a line element $d\mathbf{l}$, and $\hat{\mathbf{t}}$ is a unit vector normal to the surface element dA pointing in the direction given by the right-hand rule for the sense of integration around the contour.

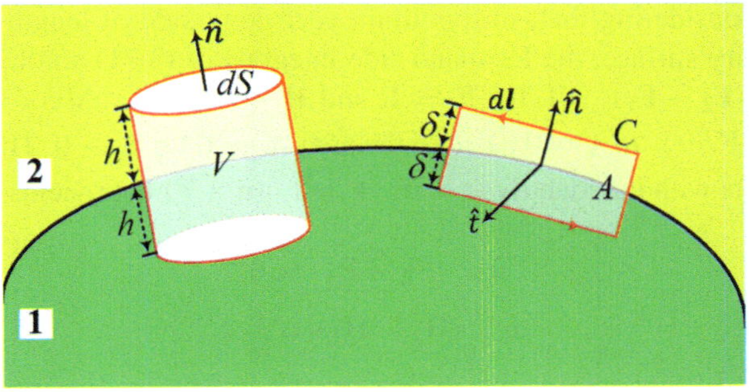

Fig. 3.6 Scheme of the boundary surface (black line) between two media. The pillbox of volume V is half in one medium and half in the other, and the unit vector $\hat{\mathbf{n}}$ is perpendicular to its top surface and pointing from medium 1 to medium 2. The rectangular contour C is oriented with its plane perpendicular to the boundary surface, so that its normal unit vector $\hat{\mathbf{t}}$ is tangent to the surface

The application of the divergence theorem (3.67) to (3.1) and (3.4) and Stokes' theorem (3.68) to (3.2) and (3.3) yields

$$\int_S \mathbf{D} \cdot \hat{\mathbf{n}} dS = \int_V \rho dV, \tag{3.69}$$

$$\int_S \mathbf{B} \cdot \hat{\mathbf{n}} dS = 0, \tag{3.70}$$

$$\int_C \mathbf{E} \cdot d\mathbf{l} = -\int_A \frac{\partial \mathbf{B}}{\partial t} \cdot \hat{\mathbf{t}} dA, \tag{3.71}$$

$$\int_C \mathbf{H} \cdot d\mathbf{l} = \int_A \left(\mathbf{J} + \frac{\partial \mathbf{D}}{\partial t} \right) \cdot \hat{\mathbf{t}} dA. \tag{3.72}$$

The Gauss' law (3.69) thus establishes that the total flux of **D** leaving the closed surface S is equal to the electric charge contained inside. The magnetic analog (3.70) yields no net flux for **B** because of the nonexistence of magnetic charges. In the limit $h \to 0$, only the top and bottom surfaces contribute to the surface integrals in (3.69) and (3.70). Considering that the two surfaces are parallel to the boundary surface and have an area ΔS each, the left-hand side integrals in (3.69) and (3.70) read $\int_S \mathbf{F} \cdot \hat{\mathbf{n}} dS = (\mathbf{F_2} - \mathbf{F_1}) \cdot \hat{\mathbf{n}} \Delta S$, for $\mathbf{F} = \mathbf{D}$ and $\mathbf{B}$. If the charge density ρ produces a surface charge density $\rho_s = (\Delta S)^{-1} \int_V \rho dV$, one obtains

$$(\mathbf{D_2} - \mathbf{D_1}) \cdot \hat{\mathbf{n}} = \rho_s, \tag{3.73}$$

$$(\mathbf{B_2} - \mathbf{B_1}) \cdot \hat{\mathbf{n}} = 0. \tag{3.74}$$

The normal components of **B** on both sides of the boundary surface are thus continuous, while the discontinuity of the corresponding ones of **D** is equal to the surface charge density.

The relationship of the components of **E** and **H** on either side of the boundary surface can also be determined by taking the limit $\delta \to 0$ of the loop C shown in Fig. 3.6. Considering that, in this limit, each long side of length Δl is parallel to the boundary surface, the left-hand side integrals of (3.71) and (3.72) reduce to $\int_C \mathbf{F} \cdot d\mathbf{l} = (\mathbf{F_2} - \mathbf{F_1}) \cdot \Delta\mathbf{l}$, for $\mathbf{F} = \mathbf{E}$ and $\mathbf{H}$, with $\Delta\mathbf{l} = \Delta l \hat{\mathbf{t}} \times \hat{\mathbf{n}}$. Further, the integral $\int_A (\partial \mathbf{F}/\partial t) \cdot \hat{\mathbf{t}} dA = 0$ because the area $A \to 0$ for $\delta \to 0$. The combination of these results with the identity $(\hat{\mathbf{t}} \times \hat{\mathbf{n}}) \cdot \mathbf{F} = \hat{\mathbf{t}} \cdot (\hat{\mathbf{n}} \times \mathbf{F})$ then yields

$$\hat{\mathbf{n}} \times (\mathbf{E_2} - \mathbf{E_1}) = 0, \tag{3.75}$$

$$\hat{\mathbf{n}} \times (\mathbf{H_2} - \mathbf{H_1}) = \mathbf{J_s}, \tag{3.76}$$

where $\mathbf{J_s} = (\Delta l)^{-1} \int_A \mathbf{J} dA$ is a surface current density flowing on the boundary surface. The tangential components of **E** across the interface are thus continuous, while the tangential components of **H** are discontinuous due to the presence of the surface current density $\mathbf{J_s}$ parallel to the boundary surface.

Considering that the normal unit vector $\hat{\mathbf{n}} = \hat{\mathbf{x}}$ is along the x axis, the boundary conditions (3.73)–(3.76) establish that the normal and tangential components of the electric and magnetic fields satisfy the following boundary conditions:

$$D_x^{(2)} - D_x^{(1)} = \rho_s, \tag{3.77}$$

$$B_x^{(2)} = B_x(1), \tag{3.78}$$

$$E_z^{(2)} = E_z^{(1)} \quad E_y^{(2)} = E_y^{(2)}, \tag{3.79}$$

$$H_z^{(2)} - H_z^{(1)} = J_{Sy} \quad H_y^{(1)} - H_y^{(2)} = J_{Sz}. \tag{3.80}$$

These boundary conditions for the four electromagnetic fields are valid for both isotropic and anisotropic media, as the constitutive relations (3.9) and (3.10) were not used to derive them. Equations (3.77)–(3.80) are thus useful to connect the solutions of the Maxwell's equations in different media and find the fields in all space.

3.4 Energy Relations

The conservation of energy of electromagnetic fields was first developed by the British physicist John Henry Poynting in 1884. This conservation law is thus often called Poynting's theorem and can be derived from Maxwell's equations. Taking into account that a differential of electrical charge dq moving with a velocity $\mathbf{v}$ in an electric field $\mathbf{E}$ and a magnetic field $\mathbf{B}$ experiences a force (Lorentz force) $d\mathbf{F} = dq\,(\mathbf{E} + \mathbf{v} \times \mathbf{B})$, and the work done by the fields is $dW = \mathbf{F} \cdot d\mathbf{x} = \mathbf{F} \cdot \mathbf{v}dt$, which yields a total work per unit time given by

$$\frac{dW}{dt} = \int \mathbf{J} \cdot \mathbf{E} dV, \tag{3.81}$$

where the definitions of the charge density $\rho = dq/dV$ and current density $\mathbf{J} = \rho\mathbf{v}$ were used. The magnetic field therefore does not work because the magnetic force is perpendicular to the velocity $[\mathbf{v} \cdot (\mathbf{v} \times \mathbf{B}) = 0]$. The power dW/dt in (3.81) represents the conversion of electromagnetic energy into mechanical or thermal energy, and it must be balanced by a decrease of energy in the electromagnetic field inside the volume V. This conservation of energy can easily be derived by multiplying the Ampere's law (3.2) by $\mathbf{E}$, which yields $\mathbf{J} \cdot \mathbf{E} = \mathbf{E} \cdot (\nabla \times \mathbf{H}) - \mathbf{E} \cdot \partial\mathbf{D}/\partial t$. After using the vector identity $\nabla \cdot (\mathbf{E} \times \mathbf{H}) = \mathbf{H} \cdot (\nabla \times \mathbf{E}) - \mathbf{E} \cdot (\nabla \times \mathbf{H})$ and the Faraday's law $\nabla \times \mathbf{E} = -\partial\mathbf{B}/\partial t$, the latter relation takes the form

$$\nabla \cdot \mathbf{S} + \frac{\partial U}{\partial t} = -\mathbf{J} \cdot \mathbf{E}, \tag{3.82}$$

where $\partial U/\partial t$ (W/m^3) is the rate of change of the internal energy density and **S** (W/m^2) is the flow of energy known as the Poynting vector. Both quantities are defined by

$$\frac{\partial U}{\partial t} = \mathbf{E} \cdot \frac{\partial \mathbf{D}}{\partial t} + \mathbf{H} \cdot \frac{\partial \mathbf{B}}{\partial t}, \tag{3.83}$$

$$\mathbf{S} = \mathbf{E} \times \mathbf{H}. \tag{3.84}$$

By inserting (3.82) into (3.81) and using the divergence theorem (3.67), one obtains

$$-\frac{d}{dt}\int U dV = \int \mathbf{S} \cdot \hat{\mathbf{n}} dA + \int \mathbf{J} \cdot \mathbf{E} dV. \tag{3.85}$$

Equations (3.82) and (3.85) are the differential and integral forms of the Poynting theorem, and their physical meaning is that the time rate of change of electromagnetic energy within a volume V is equal to the power flowing out through the boundary surface A plus the total work per unit time done by the fields on the sources in V (see Fig. 3.7).

The variation of the internal energy density U is driven by the dispersive nature of the material supporting the propagation of the fields. For instant, in a medium without dispersion, both ε and μ are real constants, and therefore (3.83) takes the form $\partial U/\partial t = \varepsilon \mathbf{E} \cdot \partial \mathbf{E}/\partial t + \mu \mathbf{H} \cdot \partial \mathbf{H}/\partial t$, which indicates that U is the sum of the energy densities of the electric and magnetic fields, as follows:

$$U = \frac{1}{2}\left(\varepsilon \left|\mathbf{E}\right|^2 + \mu \left|\mathbf{H}\right|^2\right). \tag{3.86}$$

In the presence of dispersion, the medium absorb energy and hence there is dissipation of energy. For a dispersive medium, both ε and μ are, in general, complex functions of the propagation frequency ω. For fields propagating with a single frequency (monochromatic fields), they can conveniently be expressed as

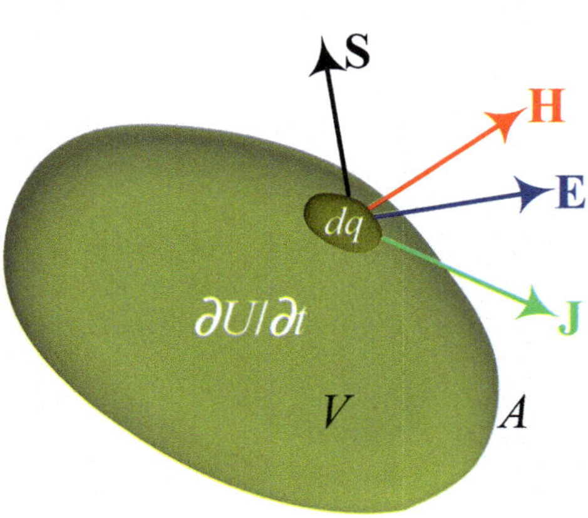

Fig. 3.7 Poynting vector generated by the propagation of the electric **E** and magnetic **H** fields. The current density **J** corresponds to the motion of the charge element dq in the volume element dV

follows: $\mathbf{F}(\mathbf{x}, t) = \text{Re}\left[\mathbf{F}(\mathbf{x}, \omega)e^{-i\omega t}\right] = 0.5\left[\mathbf{F}(\mathbf{x}, \omega)e^{-i\omega t} + \mathbf{F}^*(\mathbf{x}, \omega)e^{i\omega t}\right]$, for $\mathbf{F} = \mathbf{E}, \mathbf{D}, \mathbf{H}, \mathbf{B}$. The product $\mathbf{E} \cdot \partial\mathbf{D}/\partial t$ then becomes

$$\mathbf{E} \cdot \frac{\partial \mathbf{D}}{\partial t} = \frac{\omega}{2}\text{Im}\left[\mathbf{E}^*(\mathbf{x}, \omega) \cdot \mathbf{D}(\mathbf{x}, \omega) + \mathbf{E}(\mathbf{x}, \omega) \cdot \mathbf{D}(\mathbf{x}, \omega)e^{-2i\omega t}\right]. \tag{3.87}$$

Taking into account the relation $\mathbf{D}(\mathbf{x}, \omega) = \varepsilon(\omega)\mathbf{E}(\mathbf{x}, \omega)$ for linear–isotropic media, the average over one period $2\pi/\omega$ of the term containing the factor $e^{-2i\omega t}$ vanishes and (3.87) yields

$$\overline{\mathbf{E} \cdot \frac{\partial \mathbf{D}}{\partial t}} = \frac{\omega}{2}\text{Im}(\varepsilon(\omega)) \left|\mathbf{E}(\mathbf{x}, \omega)\right|^2, \tag{3.88}$$

where the bar denotes average with respect to time. Similarly, we can derive an analogous expression for $\mathbf{H} \cdot \partial\mathbf{B}/\partial t$, whose average value reads

$$\overline{\mathbf{H} \cdot \frac{\partial \mathbf{B}}{\partial t}} = \frac{\omega}{2}\text{Im}(\mu(\omega)) \left|\mathbf{H}(\mathbf{x}, \omega)\right|^2. \tag{3.89}$$

The average of the internal power density in (3.83) thus becomes

$$\overline{\frac{\partial U}{\partial t}} = \frac{\omega}{2}\left[\text{Im}(\varepsilon(\omega)) \left|\mathbf{E}(\mathbf{x}, \omega)\right|^2 + \text{Im}(\mu(\omega)) \left|\mathbf{H}(\mathbf{x}, \omega)\right|^2\right]. \tag{3.90}$$

The absorption (dissipation) of energy related to the electric and magnetic fields is thus determined by the imaginary parts of ε and μ, respectively. Following a similar procedure to determine the average of the Poynting vector $\mathbf{S} = 2^{-2}\left[\mathbf{E}(\mathbf{x}, \omega)e^{-i\omega t} + \mathbf{E}^*(\mathbf{x}, \omega)e^{i\omega t}\right] \times \left[\mathbf{H}(\mathbf{x}, \omega)e^{-i\omega t} + \mathbf{H}^*(\mathbf{x}, \omega)e^{i\omega t}\right]$ and thermal term $Q = \mathbf{J} \cdot \mathbf{E} = 2^{-2}\left[\mathbf{J}(\mathbf{x}, \omega)e^{-i\omega t} + \mathbf{J}^*(\mathbf{x}, \omega)e^{i\omega t}\right] \cdot \left[\mathbf{E}(\mathbf{x}, \omega)e^{-i\omega t} + \mathbf{E}^*(\mathbf{x}, \omega)e^{i\omega t}\right]$, one obtains

$$\overline{\mathbf{S}} = \frac{1}{2}\text{Re}\left[\mathbf{E}(\mathbf{x}, \omega) \times \mathbf{H}^*(\mathbf{x}, \omega)\right], \tag{3.91}$$

$$\overline{Q} = \frac{1}{2}\text{Re}\left[\mathbf{J}(\mathbf{x}, \omega) \cdot \mathbf{E}^*(\mathbf{x}, \omega)\right] = \frac{\omega}{2}\text{Im}\left(\varepsilon(\omega)\right) \left|\mathbf{E}(\mathbf{x}, \omega)\right|^2, \tag{3.92}$$

where we have used the Ohm law (3.54) and the relation (3.56) between the permittivity and electric conductivity. The average quantities in (3.90)–(3.92) then allow to write the Poynting theorem (3.82) in the frequency domain, as follows:

$$\nabla \cdot \overline{\mathbf{S}} + \overline{\frac{\partial U}{\partial t}} = -\overline{Q}. \tag{3.93}$$

For non-monochromatic fields propagating in a wide spectrum of frequencies, we can also derive simple formulas for the components of the Poynting theorem in the frequency domain by expressing the field $\mathbf{F}(\mathbf{x}, t)$ in terms of its Fourier transform $\mathbf{F}(\mathbf{x}, \omega)$ defined in (3.6). The Fourier's integrals of $\mathbf{E}(\mathbf{x}, t) = \int_{-\infty}^{\infty} \mathbf{E}(\mathbf{x}, \omega')e^{-i\omega' t}d\omega'$ and $\partial\mathbf{D}(\mathbf{x}, t)/\partial t = -i\int_{-\infty}^{\infty} \omega\mathbf{D}(\mathbf{x}, \omega)e^{-i\omega t}d\omega$ then allow to write their product as the double integral

$$\mathbf{E}\cdot\frac{\partial\mathbf{D}}{\partial t} = -i\int_{-\infty}^{\infty} d\omega \int_{-\infty}^{\infty} \omega\varepsilon(\omega)\mathbf{E}(\mathbf{x}, \omega)\cdot\mathbf{E}(\mathbf{x}, \omega')e^{-i(\omega+\omega')t}, \tag{3.94}$$

where we have used the relation $\mathbf{D}(\mathbf{x}, \omega) = \varepsilon(\omega)\mathbf{E}(\mathbf{x}, \omega)$ for linear and isotropic media. The integration of (3.94) over the whole existence of the fields ($-\infty < t < \infty$) can easily be evaluated by means of the Dirac delta function $\delta(\omega + \omega') = (2\pi)^{-1}\int_{-\infty}^{\infty} e^{-i(\omega+\omega')t}dt$, which yields

$$\left\langle\mathbf{E}\cdot\frac{\partial\mathbf{D}}{\partial t}\right\rangle = \int_{-\infty}^{\infty} \mathbf{E}\cdot\frac{\partial\mathbf{D}}{\partial t}dt = -2\pi i\int_{-\infty}^{\infty} \omega\varepsilon(\omega)\mathbf{E}(\mathbf{x}, \omega)\cdot\mathbf{E}(\mathbf{x}, -\omega)d\omega. \tag{3.95}$$

Taking into account that $\mathbf{E}(\mathbf{x}, -\omega) = \mathbf{E}^*(\mathbf{x}, \omega)$, the product $\mathbf{E}(\mathbf{x}, \omega)\cdot\mathbf{E}(\mathbf{x}, -\omega) = |\mathbf{E}(\mathbf{x}, \omega)|^2$ is an even function of frequency ω. Further, the property $\varepsilon(-\omega) = \varepsilon^*(\omega)$ indicates that the real and imaginary parts of $\varepsilon(\omega) = \varepsilon_R(\omega) + i\varepsilon_I(\omega)$ are even $[\varepsilon_R(-\omega) = \varepsilon_R(\omega)]$ and odd $[\varepsilon_I(-\omega) = -\varepsilon_I(\omega)]$ functions of ω. The integrand containing the term $\varepsilon_R(\omega)$ $[\varepsilon_I(\omega)]$ is thus an odd (even) function of ω, and therefore its integration vanishes (is equal to twice the integral from 0 to ∞). Equation (3.95) then takes the form

$$\left\langle\mathbf{E}\cdot\frac{\partial\mathbf{D}}{\partial t}\right\rangle = 4\pi\int_{0}^{\infty} \omega\mathrm{Im}(\varepsilon(\omega))\,|\mathbf{E}(\mathbf{x}, \omega)|^2\,d\omega. \tag{3.96}$$

With a similar expression for the product $\mathbf{H}\cdot\partial\mathbf{B}/\partial t$ of magnetic fields, we finally have

$$\left\langle\frac{\partial U}{\partial t}\right\rangle = 4\pi\int_{0}^{\infty} \omega\left[\mathrm{Im}(\varepsilon(\omega))\,|\mathbf{E}(\mathbf{x}, \omega)|^2 + \mathrm{Im}(\mu(\omega))\,|\mathbf{H}(\mathbf{x}, \omega)|^2\right]d\omega. \tag{3.97}$$

The variation of the internal electromagnetic energy is thus driven by the frequency spectrum of the imaginary parts of ε and μ. By conservation of energy, this variation is accompanied by the generation of heat ($Q > 0$) within the medium and a power flowing outside it ($|\mathbf{S}| > 0$). After expression $Q = \mathbf{J}\cdot\mathbf{E}$ in terms of a double integral, as we did in (3.94), and integrating over the whole duration of the fields, one obtains

$$\langle Q\rangle = \int_{-\infty}^{\infty} Q dt = 2\pi\int_{-\infty}^{\infty} \sigma(\omega)\,|\mathbf{E}(\mathbf{x}, \omega)|^2\,d\omega, \tag{3.98}$$

where the electrical conductivity $\sigma(\omega) = \sigma_R(\omega) + i\sigma_I(\omega)$ satisfies the condition $\sigma(-\omega) = \sigma^*(\omega)$ due to its direct correlation with the permittivity in (3.56). We thus obtain $\sigma_R(-\omega) = \sigma_R(\omega)$ and $\sigma_I(-\omega) = -\sigma_I(\omega)$, which indicate that the integral in (3.98) over σ_I vanishes, while the one over σ_R can be replaced by two times the integral from 0 to ∞. Further, considering that $\sigma_R = \omega\varepsilon_I$ (in absolute units of permittivity), as established by (3.56), (3.98) reduces to

$$\langle Q \rangle = 4\pi \int_0^{\infty} \omega \mathrm{Im}(\varepsilon(\omega)) \left|\mathbf{E}(\mathbf{x}, \omega)\right|^2 d\omega. \tag{3.99}$$

The generated heat thus represents the absorptive dissipation in the medium that increases with the frequency spectrum of the imaginary part of ε, as is the case of the internal energy variation in (3.97). Since $Q \geq 0$, (3.99) determines that the imaginary part of the permittivity is always positive or zero ($\mathrm{Im}(\varepsilon) \geq 0$). Similarly, the integrated Poynting vector is

$$\langle \mathbf{S} \rangle = \int_{-\infty}^{\infty} S dt = 2\pi \int_{-\infty}^{\infty} \mathbf{E}(\mathbf{x}, \omega) \times \mathbf{H}^*(\mathbf{x}, \omega) d\omega. \tag{3.100}$$

The reality condition $\langle \mathbf{S} \rangle = \langle \mathbf{S} \rangle^*$ and the sum of (3.100) with its complex conjugate indicate that the integrand in (3.100) can be replaced by $\mathrm{Re}[\mathbf{E}(\mathbf{x}, \omega) \times \mathbf{H}^*(\mathbf{x}, \omega)] = \mathrm{Re}(\mathbf{E}(\mathbf{x}, \omega)) \times \mathrm{Re}(\mathbf{H}(\mathbf{x}, \omega)) + \mathrm{Im}(\mathbf{E}(\mathbf{x}, \omega)) \times \mathrm{Im}(\mathbf{H}(\mathbf{x}, \omega))$, which is an even function of ω, since $\mathbf{E}(\mathbf{x}, -\omega) = \mathbf{E}^*(\mathbf{x}, \omega)$ and $\mathbf{H}(\mathbf{x}, -\omega) = \mathbf{H}^*(\mathbf{x}, \omega)$. We thus have

$$\langle \mathbf{S} \rangle = 4\pi \int_0^{\infty} \mathrm{Re}\left[\mathbf{E}(\mathbf{x}, \omega) \times \mathbf{H}^*(\mathbf{x}, \omega)\right] d\omega, \tag{3.101}$$

which represents the extension of (3.91) for non-monochromatic fields. In terms of the integrated quantities in (3.97), (3.99), and (3.101), the Poynting theorem (3.82) becomes

$$\nabla \cdot \langle \mathbf{S} \rangle + \left\langle \frac{\partial U}{\partial t} \right\rangle = -\langle Q \rangle. \tag{3.102}$$

Equation (3.102) states the local conservation of electromagnetic energy in realistic situations, for which the energy flowing out of the medium ($\nabla \cdot \langle \mathbf{S} \rangle \neq 0$) and the energy losses for heating it ($\langle Q \rangle \geq 0$) lead to the decay of the energy in the fields.

3.5 Equations for the Fields' Components

Taking into account that the solution of nonhomogeneous differential equations can be built up from the solution of their homogeneous counterparts, the simplest and most fundamental solution of the Maxwell's equations appears in uniform isotropic–linear media without internal sources ($\rho = 0 = J$). In these media, the electrical permittivity ε and magnetic permeability μ are, in general, complex functions of frequency, and the Maxwell's equations (3.7) and (3.8) read

$$\nabla \times \mathbf{H} + i\omega\varepsilon\mathbf{E} = 0, \tag{3.103}$$

$$\nabla \times \mathbf{E} - i\omega\mu\mathbf{H} = 0. \tag{3.104}$$

Considering that the fields $\mathbf{E}(\mathbf{x}, \omega)$ and $\mathbf{H}(\mathbf{x}, \omega)$ propagate as waves along a straight line (i.e., the z axis), they can be written as $\mathbf{F}(\mathbf{x}, \omega) = \mathbf{F}_0(u, v, \omega)e^{i\beta z}$, where β is the propagation wave vector and $(u, v) = (x, y)$ or (r, θ) are the orthogonal Cartesian or cylindrical coordinates, respectively. In both coordinate systems, the gradient operator can be written as $\nabla = \nabla_\perp + \hat{z}\partial/\partial z$, which yields $\nabla \times \mathbf{F} = \left(\nabla_\perp \times \mathbf{F}_0 + i\beta\hat{z} \times \mathbf{F}_0\right) e^{i\beta z}$. By inserting this expression into (3.103) and (3.104), one obtains

$$\omega\varepsilon\mathbf{E}_0 + \beta\hat{z} \times \mathbf{H}_0 = i\nabla_\perp \times \mathbf{H}_0, \tag{3.105}$$

$$\beta\hat{z} \times \mathbf{E}_0 - \omega\mu\mathbf{H}_0 = i\nabla_\perp \times \mathbf{E}_0. \tag{3.106}$$

The scalar components of both fields can now be determined for a particular coordinate system.

3.5.1 Cartesian Coordinates

In Cartesian coordinates (x, y, z), the components of $\mathbf{F}_0 = F_x\hat{x} + F_y\hat{y} + F_z\hat{z}$ yield $\hat{z} \times \mathbf{F}_0 = F_x\hat{y} - F_y\hat{x}$ and $\nabla_\perp \times \mathbf{F}_0 = \hat{x}\partial F_z/\partial y - \hat{y}\partial F_z/\partial x + \hat{z}\left(\partial F_y/\partial x - \partial F_x/\partial y\right)$. These two relations transform the vector equations (3.105) and (3.106) into the following six scalar equations:

$$\omega\varepsilon E_x - \beta H_y = i\frac{\partial H_z}{\partial y}, \tag{3.107}$$

$$\omega\varepsilon E_y + \beta H_x = -i\frac{\partial H_z}{\partial x}, \tag{3.108}$$

$$\omega\varepsilon E_z = i\left(\frac{\partial H_y}{\partial x} - \frac{\partial H_x}{\partial y}\right), \tag{3.109}$$

$$\beta E_y + \omega\mu H_x = -i\frac{\partial E_z}{\partial y}, \tag{3.110}$$

$$\beta E_x - \omega\mu H_y = -i\frac{\partial E_z}{\partial x}, \tag{3.111}$$

$$\omega\mu H_z = -i\left(\frac{\partial E_y}{\partial x} - \frac{\partial E_x}{\partial y}\right). \tag{3.112}$$

After solving the pair of Eqs. (3.107) and (3.111) for E_x and H_y and (3.108) and (3.110) for E_y and H_x, we have

$$E_x = -\frac{i}{p^2}\left(\beta\frac{\partial E_z}{\partial x} + \omega\mu\frac{\partial H_z}{\partial y}\right), \tag{3.113}$$

$$E_y = -\frac{i}{p^2}\left(\beta\frac{\partial E_z}{\partial y} - \omega\mu\frac{\partial H_z}{\partial x}\right), \tag{3.114}$$

$$H_x = -\frac{i}{p^2}\left(\beta\frac{\partial H_z}{\partial x} - \omega\varepsilon\frac{\partial E_z}{\partial y}\right), \tag{3.115}$$

$$H_y = -\frac{i}{p^2}\left(\beta\frac{\partial H_z}{\partial y} + \omega\varepsilon\frac{\partial E_z}{\partial x}\right), \tag{3.116}$$

where $p^2 = \beta^2 - \varepsilon\mu\omega^2$. All the transverse field components can therefore be determined from the longitudinal axial ones E_z and H_z. Inserting (3.115) and (3.116) into (3.109) and (3.113) and (3.114) into (3.112), we find that both of these latter components satisfy the Helmholtz equation

$$\left(\nabla_\perp - p^2\right)\begin{pmatrix} E_z \\ H_z \end{pmatrix} = 0, \tag{3.117}$$

with $\nabla_\perp^2 = \partial^2/\partial x^2 + \partial^2/\partial y^2$ being the transverse Laplacian in Cartesian coordinates.

3.5.2 Cylindrical Coordinates

In cylindrical coordinates (r, θ, z), we have $\mathbf{F}_0 = F_r\hat{r} + F_\theta\hat{\theta} + F_z\hat{z}$, which yields $\hat{z} \times \mathbf{F}_0 = F_r\hat{\theta} - F_\theta\hat{r}$ and $\nabla_\perp \times \mathbf{F}_0 = (\hat{r}/r)\partial F_z/\partial\theta - \hat{\theta}\partial F_z/\partial r + (\hat{z}/r)\left[\partial(rF_\theta)/\partial r - \partial F_r/\partial\theta\right]$. By substituting these relations into (3.105) and (3.106), we obtain

$$\omega\varepsilon E_r - \beta H_\theta = \frac{i}{r}\frac{\partial H_z}{\partial\theta}, \tag{3.118}$$

$$\omega\varepsilon E_\theta + \beta H_r = -i\frac{\partial H_z}{\partial r}, \tag{3.119}$$

$$\omega\varepsilon E_z = \frac{i}{r}\left(\frac{\partial(rH_\theta)}{\partial r} - \frac{\partial H_r}{\partial\theta}\right), \tag{3.120}$$

$$\beta E_\theta + \omega\mu H_r = -\frac{i}{r}\frac{\partial E_z}{\partial\theta}, \tag{3.121}$$

$$\beta E_r - \omega\mu H_\theta = -i\frac{\partial E_z}{\partial r}, \tag{3.122}$$

$$\omega\mu H_z = -\frac{i}{r}\left(\frac{\partial(rE_\theta)}{\partial r} - \frac{\partial E_r}{\partial\theta}\right). \tag{3.123}$$

The solution of the system of Eqs. (3.118)–(3.123) for the transverse field components as functions of the longitudinal ones, as we did in Cartesian coordinates, yields

$$E_r = -\frac{i}{p^2}\left(\beta\frac{\partial E_z}{\partial r} + \frac{\omega\mu}{r}\frac{\partial H_z}{\partial \theta}\right), \tag{3.124}$$

$$E_\theta = -\frac{i}{p^2}\left(\frac{\beta}{r}\frac{\partial E_z}{\partial \theta} - \omega\mu\frac{\partial H_z}{\partial r}\right), \tag{3.125}$$

$$H_r = -\frac{i}{p^2}\left(\beta\frac{\partial H_z}{\partial r} - \frac{\omega\varepsilon}{r}\frac{\partial E_z}{\partial \theta}\right), \tag{3.126}$$

$$H_\theta = -\frac{i}{p^2}\left(\frac{\beta}{r}\frac{\partial H_z}{\partial \theta} + \omega\varepsilon\frac{\partial E_z}{\partial r}\right), \tag{3.127}$$

where E_z and H_z satisfy (3.117) with the transverse Laplacian $\nabla_\perp^2 = (1/r)\partial(r\partial/\partial r)/\partial r + (1/r^2)\partial^2/\partial\theta^2$ in cylindrical coordinates. Since (3.117) can be solved by the method of separation of variables, Eqs. (3.113)–(3.116) and (3.124)–(3.127) are the analytical solutions for the field components in Cartesian and cylindrical coordinates. As the transverse components are determined by a linear combination of the derivatives of the longitudinal ones, the solution with $E_z \neq 0 \neq H_z$ is determined by the linear superposition of the transverse electric (TE: $E_z = 0 \neq H_z$) and transverse magnetic (TM: $E_z \neq 0 = H_z$) waves. Even though the nonvanishing longitudinal components of these two basic wave types satisfy the same equation (3.117), they fulfill different boundary conditions, as established by (3.75) and (3.76).

Taking into account that $\exp(i\beta z)\exp(-i\beta^* z) = \exp[-2\mathrm{Im}(\beta)z]$, the spectra of the heat source and Poynting vector defined in (3.91) and (3.92) take the form

$$\overline{Q} = \frac{\omega\varepsilon_0}{2}\mathrm{Im}\,(\varepsilon)\,|\mathbf{E}_0|^2\, e^{-z/\Lambda}, \tag{3.128}$$

$$\overline{\mathbf{S}} = \frac{1}{2}\mathrm{Re}\left[\mathbf{E}_0 \times \mathbf{H}_0^*\right] e^{-z/\Lambda}, \tag{3.129}$$

where Λ is the distance for which the wave intensity decays a factor e^{-1} defined by

$$\Lambda = \frac{1}{2\mathrm{Im}(\beta)}. \tag{3.130}$$

The propagation length of the wave energy is thus totally determined by the imaginary part of the wave vector along the propagation direction (z axis).

References

1. Lorentz, H.A.: The theory of electrons and its applications to the phenomena of light and radiant heat. B. G. Teubner, Leipzig (1909)
2. Drude, P.: Zur elektronentheorie der metalle. Ann. Phys. **306**, 566–613 (1900)
3. Borenstein, M., Lamb, W.E.: Classical laser. Phys. Rev. A **5**, 1298–1311 (1972)

Chapter 4
Dispersion Relations and Propagation Parameters of Polaritons

4.1 Planar Structures

Considering that the electric and magnetic fields are uniformly excited along the y axis and propagate along the z axis of an infinite medium in both the z and y directions, the field components become independent of the y coordinate. Under this condition, (3.117) reduces to $\partial^2 E_z/\partial x^2 - p^2 E_z = 0$, whose two solutions are $E_z = e^{\pm px}$. According to (3.113)–(3.116), the nonzero field components (E_x, E_z, and H_y) of the TM waves ($H_z = 0$) are then given by

$$E_x = -\frac{i\beta}{p^2}\frac{\partial E_z}{\partial x}, \tag{4.1}$$

$$H_y = -\frac{i\omega\varepsilon_0\varepsilon}{p^2}\frac{\partial E_z}{\partial x}, \tag{4.2}$$

$$E_z = Ae^{px} + Be^{-px}, \tag{4.3}$$

where A and B are constants determined by the boundary conditions for a particular problem and

$$p^2 = \beta^2 - \varepsilon k_0^2, \tag{4.4}$$

with $k = \omega/c$ and $c = 1/\sqrt{\varepsilon_0\mu_0}$ being the wave vector and speed of light in vacuum, respectively. To conveniently write p in terms of the wave vectors β and k_0, we have redefined ε in (4.2) and (4.4) as the relative (dimensionless) permittivity $\varepsilon/\varepsilon_0$. For now on, for the sake of simplicity, we are going to use this redefinition ($\varepsilon \to \varepsilon/\varepsilon_0$) in all equations.

As the TM waves have two components for the electrical field, (3.128) establishes that the spectral thermal energy generated by them inside the medium of relative permittivity ε is given by

S. Volz, J. Ordonez-Miranda, *Heat Transport Driven by Surface Electromagnetic Waves*, Mechanical Engineering Series, https://doi.org/10.1007/978-3-031-66024-5_4

$$\overline{Q} = \frac{\omega\varepsilon_0}{2}\text{Im}\,(\varepsilon)\left(|E_x|^2 + |E_z|^2\right)e^{-z/\Lambda}. \tag{4.5}$$

Taking into account that $\mathbf{E}_0 \times \mathbf{H}_0^* = \left(E_x\hat{x} + E_z\hat{z}\right) \times H_y^*\hat{y} = \left(E_x\hat{z} - E_z\hat{x}\right)H_y^*$, the substitution of (4.1) and (4.2) into (3.129) yields the following components for the spectral Poynting vector of TM waves

$$S_z = \frac{\omega\varepsilon_0}{2\,|p|^4}\text{Re}\left(\beta\varepsilon^*\right)\left|\frac{\partial E_z}{\partial x}\right|^2 e^{-z/\Lambda}, \tag{4.6}$$

$$S_x = -\frac{\omega\varepsilon_0}{2}\text{Im}\left(\frac{\varepsilon E_z^*}{p^2}\frac{\partial E_z}{\partial x}\right)e^{-z/\Lambda}. \tag{4.7}$$

On the other hand, the nonvanishing field components (H_x, H_z, and E_y) of the TE waves ($E_z = 0$) predicted by (3.113)–(3.117) are

$$H_x = -\frac{i\beta}{p^2}\frac{\partial H_z}{\partial x}, \tag{4.8}$$

$$E_y = \frac{i\omega\mu_0}{p^2}\frac{\partial H_z}{\partial x}, \tag{4.9}$$

$$H_z = Ce^{px} + De^{-px}, \tag{4.10}$$

where C and D are constants. For both TM and TE waves (see Fig. 4.1), the parameter p represents a complex wave vector driving the propagation of the fields along the transverse x direction, as shown by (4.3) and (4.10). In contrast to the TM waves, the TE ones have only one component for the electrical field, and therefore their associated thermal source in (3.128) reduces to

$$\overline{Q} = \frac{\omega\varepsilon_0}{2}\text{Im}\,(\varepsilon)\left|E_y\right|^2 e^{-z/\Lambda}. \tag{4.11}$$

For TE waves, the product $\mathbf{E}_0 \times \mathbf{H}_0^* = E_y\hat{y} \times \left(H_x^*\hat{x} + H_z^*\hat{z}\right) = E_y\left(-H_x^*\hat{z} + H_z^*\hat{x}\right)$, and hence the combination of (4.8) and (4.9) with (3.129), yields the following components for the Poynting vector:

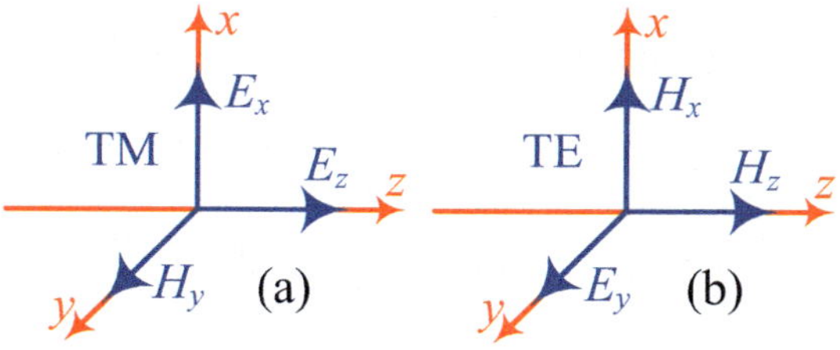

Fig. 4.1 Field components of the (**a**) transverse magnetic (TM) and (**b**) transverse electric (TE) waves

$$S_z = \frac{\omega\mu_0}{2\,|p|^4}\mathrm{Re}\,(\beta)\left|\frac{\partial H_z}{\partial x}\right|^2 e^{-z/\Lambda}, \tag{4.12}$$

$$S_x = -\frac{\omega\mu_0}{2}\mathrm{Im}\left(\frac{H_z^*}{p^2}\frac{\partial H_z}{\partial x}\right)e^{-z/\Lambda}. \tag{4.13}$$

The two components of the Poynting vector of TM or TE waves can thus be determined after calculating the longitudinal component of their electric or magnetic field, respectively. According to (3.79) and (3.80), the tangential field components in (4.1)–(4.3) and (4.8)–(4.10) are subjected to the following boundary conditions at the interface of two media:

$$E_z^{(2)} = E_z^{(1)} \qquad H_y^{(2)} = H_y^{(1)}, \tag{4.14}$$

$$H_z^{(2)} = H_z^{(1)} \qquad E_y^{(2)} = E_y^{(1)}. \tag{4.15}$$

We are now going to apply these boundary conditions to study the propagation of TM and TE waves supported by particular planar structures.

4.2 Polariton Propagation Along a Single Interface

Here we consider that two semi-infinite media spanning the x axis support the propagation of surface polaritons along their interface $x = 0$, as shown in Fig. 4.2. The relative permittivity of medium 1 ($x < 0$) and 2 ($x > 0$) is, respectively, ε_1 and ε_2, and therefore the cross-plane wave vector inside them is given by $p_n^2 = \beta^2 - \varepsilon_n k_0^2$, as established by (4.4) for $n = 1$ and 2.

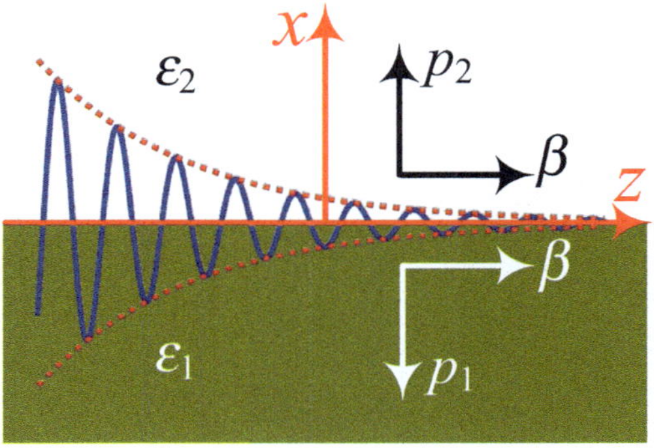

Fig. 4.2 Propagation of polaritons along the interface $x = 0$ of two semi-infinite media of permittivities ε_1 ($x < 0$) and ε_2 ($x > 0$). The in-plane oscillations (blue line) and the cross-plane decay (red dashed lines) represent the propagative and evanescent behavior of the polariton fields traveling along the z and x directions

4.2.1 TE Polarization

Taking into account that the physical solution for the field components must vanish or remind finite for $x \to \pm\infty$, the TE field components predicted by (4.8)–(4.10) in medium $n = 1$ and 2 are

$$H_z^{(1)} = Ce^{p_1 x}, \quad H_x^{(1)} = -\frac{i\beta}{p_1} H_z^{(1)}, \quad E_y^{(1)} = \frac{i\omega\mu_0}{p_1} H_z^{(1)}, \tag{4.16}$$

$$H_z^{(2)} = De^{-p_2 x}, \quad H_x^{(2)} = \frac{i\beta}{p_2} H_z^{(2)}, \quad E_y^{(2)} = -\frac{i\omega\mu_0}{p_2} H_z^{(2)}, \tag{4.17}$$

where we have considered that $\mathrm{Re}(p_n) > 0$ to ensure the exponential decay away from the interface $x = 0$. This condition ($\mathrm{Re}(p_n) > 0$) guarantees the existence of surface polaritons, which exist when all field components decay as they travel away from $x = 0$. The continuity of the tangential components H_z and E_y at the boundary surface $x = 0$ yields $C = D$ and

$$p_1 + p_2 = 0. \tag{4.18}$$

Equation (4.18) does not satisfy the physical condition $\mathrm{Re}(p_n) > 0$, and therefore the polariton TE waves do not exist ($C = D = 0$) along a single interface.

4.2.2 TM Polarization

For the TM waves, the field components predicted by (4.1)–(4.3) are

$$E_z^{(1)} = Ae^{p_1 x}, \quad E_x^{(1)} = -\frac{i\beta}{p_1} E_z^{(1)}, \quad H_y^{(1)} = -\frac{i\omega\varepsilon_0\varepsilon_1}{p_1} E_z^{(1)}, \tag{4.19}$$

$$E_z^{(2)} = Be^{-p_2 x}, \quad E_x^{(2)} = \frac{i\beta}{p_2} E_z^{(2)}, \quad H_y^{(2)} = \frac{i\omega\varepsilon_0\varepsilon_2}{p_2} E_z^{(2)}. \tag{4.20}$$

Matching the tangential field components E_z and H_y at the boundary surface $x = 0$, one obtains $A = B$ and

$$\frac{p_1}{\varepsilon_1} + \frac{p_2}{\varepsilon_2} = 0. \tag{4.21}$$

Using the relation $p_n = \sqrt{\beta^2 - \varepsilon_n k_0^2}$, the solution of (4.21) for the in-plane (β) and cross-plane (p_n) wave vectors is

$$\beta = k_0 \sqrt{\frac{\varepsilon_1 \varepsilon_2}{\varepsilon_1 + \varepsilon_2}}, \tag{4.22}$$

$$p_1 = -\frac{k_0 \varepsilon_1}{\sqrt{-\varepsilon_1 - \varepsilon_2}}, \tag{4.23}$$

$$p_2 = \frac{k_0 \varepsilon_2}{\sqrt{-\varepsilon_1 - \varepsilon_2}}, \tag{4.24}$$

where we have assumed that the negative sign in front of the right-hand side of (4.23) ensures the physical condition $\mathrm{Re}(p_1) > 0$. Otherwise, according to (4.21), we have the free choice to assign that negative sign to (4.24), provided that $\mathrm{Re}(p_2) > 0$. For nonabsorbing media with real permittivities ε_1 and ε_2, the condition $\mathrm{Re}(p_n) > 0$ and (4.23) and (4.24) indicate that the TM waves exist for $\varepsilon_1 < 0$ and $\varepsilon_2 > 0$, such that $-\varepsilon_1 - \varepsilon_2 > 0$. These constraints thus establish that the TM waves are surface polaritons for

$$\varepsilon_1 < -\varepsilon_2 < 0. \tag{4.25}$$

Under this condition, the in-plane wave vector β in (4.22) is real and positive, which confirms the propagation of surface polaritons along the media interface. For an absorbing medium 1 (complex $\varepsilon_1 = \varepsilon_R + i\varepsilon_I$) and a nonabsorbing medium 2 (real $\varepsilon_2 > 0$), on the other hand, β and p_n become complex, and their real and imaginary parts can conveniently be determined by rewriting (4.22)–(4.24) as follows:

$$\beta = \frac{k_0 \sqrt{\varepsilon_2}}{|\varepsilon_1 + \varepsilon_2|} \sqrt{|\varepsilon_1|^2 + \varepsilon_1 \varepsilon_2}, \tag{4.26}$$

$$p_1 = -\frac{k_0 \varepsilon_1}{|\varepsilon_1 + \varepsilon_2|} \sqrt{-\varepsilon_1^* - \varepsilon_2}, \tag{4.27}$$

$$p_2 = \frac{k_0 \varepsilon_2}{|\varepsilon_1 + \varepsilon_2|} \sqrt{-\varepsilon_1^* - \varepsilon_2}, \tag{4.28}$$

where $|Z|$ and Z^* stand for the amplitude and complex conjugate of Z, respectively. Since the square root of any complex number $a + ib$ can be written as $\sqrt{a + ib} = (\chi_+ + i\chi_-)/\sqrt{2}$, with $\chi_\pm = \sqrt{|a + ib| \pm a}$ for $b > 0$, the real and imaginary parts of $\beta = \beta_R + i\beta_I$ and $p_n = p_{nR} + ip_{nI}$ predicted by (4.26)–(4.28) are

$$\beta_R = \frac{k_0 \sqrt{\varepsilon_2}}{\sqrt{2}\,|\varepsilon_1 + \varepsilon_2|} \sqrt{|\varepsilon_1| \left(|\varepsilon_1 + \varepsilon_2| + |\varepsilon_1|\right) + \varepsilon_R \varepsilon_2}, \tag{4.29}$$

$$\beta_I = \frac{k_0 \sqrt{\varepsilon_2}}{\sqrt{2}\,|\varepsilon_1 + \varepsilon_2|} \sqrt{|\varepsilon_1| \left(|\varepsilon_1 + \varepsilon_2| - |\varepsilon_1|\right) - \varepsilon_R \varepsilon_2}, \tag{4.30}$$

$$p_{1R} = \frac{k_0}{\sqrt{2}\,|\varepsilon_1 + \varepsilon_2|} \left(\varepsilon_I \sqrt{|\varepsilon_1 + \varepsilon_2| + \varepsilon_R + \varepsilon_2} - \varepsilon_R \sqrt{|\varepsilon_1 + \varepsilon_2| - \varepsilon_R - \varepsilon_2}\right), \tag{4.31}$$

$$p_{1I} = \frac{-k_0}{\sqrt{2}\,|\varepsilon_1 + \varepsilon_2|} \left(\varepsilon_R \sqrt{|\varepsilon_1 + \varepsilon_2| + \varepsilon_R + \varepsilon_2} + \varepsilon_I \sqrt{|\varepsilon_1 + \varepsilon_2| - \varepsilon_R - \varepsilon_2}\right), \tag{4.32}$$

$$p_{2R} = \frac{k_0 \varepsilon_2}{\sqrt{2}\,|\varepsilon_1 + \varepsilon_2|}\sqrt{|\varepsilon_1 + \varepsilon_2| - \varepsilon_R - \varepsilon_2}, \quad (4.33)$$

$$p_{2I} = \frac{k_0 \varepsilon_2}{\sqrt{2}\,|\varepsilon_1 + \varepsilon_2|}\sqrt{|\varepsilon_1 + \varepsilon_2| + \varepsilon_R + \varepsilon_2}. \quad (4.34)$$

For $\varepsilon_I = 0$, the imaginary parts of the wave vectors vanish ($\beta_I = p_{1I} = p_{2I} = 0$), while their real parts reduce to (4.22)–(4.24), for waves satisfying the condition (4.25), as expected. By contrast, for $\varepsilon_I > 0$, the real parts of the wave vectors are all positive ($\beta_R > 0$, $p_{1R} > 0$, and $p_{2R} > 0$), for any relation between the real and imaginary parts of the permittivity ε_1. While these positivity conditions are evident for β_R and p_{2R} in (4.29) and (4.33), the one for p_{1R} can be demonstrated by using (4.31) to rewrite the condition $p_{1R} > 0$ as follows: $\varepsilon_I\sqrt{|\varepsilon_1 + \varepsilon_2| + \varepsilon_R + \varepsilon_2} > \varepsilon_R\sqrt{|\varepsilon_1 + \varepsilon_2| - \varepsilon_R - \varepsilon_2}$. Given that ε_I is always positive, this inequality is evidently valid for $\varepsilon_R < 0$. This validity also holds for $\varepsilon_R > 0$, as the squaring of the inequality leads to $\varepsilon_R^2 + \varepsilon_I^2 + 2\varepsilon_R\varepsilon_2 > 0$, which is always true, for the considered nonabsorbing medium with $\varepsilon_2 > 0$. The propagation of TM waves along the interface of absorbing/nonabsorbing semi-infinite media ($\beta_R > 0$) is accompanied by their cross-plane attenuation ($p_{1R} > 0$ and $p_{2R} > 0$) for all frequencies. This fact indicates that this latter interface supports the propagation of polaritons in a broader frequency range than the former one, which maintains the existence of polaritons for frequencies satisfying the condition (4.25) only. Furthermore, the in-plane propagation along the interface of absorbing/nonabsorbing media is driven by the attenuation factor $\beta_I > 0$, which increases with ε_I. A longer propagation length $\Lambda = 1/2\beta_I$ (see 3.130) is therefore expected for smaller absorption (lower ε_I), as expected.

The role of the energy absorption on the propagation parameters can explicitly be seen in the product of the real and imaginary parts of each wave vector. Based on (4.29)–(4.34), we obtain

$$\beta_R\beta_I = p_{2R}p_{2I} = \frac{\varepsilon_I}{2}\left(\frac{k_0\varepsilon_2}{|\varepsilon_1 + \varepsilon_2|}\right)^2, \quad (4.35)$$

$$p_{1R}p_{1I} = -\frac{\varepsilon_I}{2}k_0^2\left[1 - \left(\frac{\varepsilon_2}{|\varepsilon_1 + \varepsilon_2|}\right)^2\right]. \quad (4.36)$$

The product of the two parts of each wave vector is thus driven by its linear dependence on ε_I. This correlation arises from either the real or the imaginary part of each wave vector, as both can be proportional to ε_I, depending on the values of ε_R. This behavior can explicitly be seen in the limit of low absorption ($\varepsilon_I \ll |\varepsilon_R|; |\varepsilon_R + \varepsilon_2|$). In this case, the binomial series expansions of (4.29)–(4.30), for an approximation up to ε_I^2, yield

$$\beta_R = k_0 \sqrt{\frac{\varepsilon_R \varepsilon_2}{\varepsilon_R + \varepsilon_2}} \left[1 + \frac{\varepsilon_2}{8} (\varepsilon_2 + 4\varepsilon_R) \left(\frac{\varepsilon_I}{\varepsilon_R(\varepsilon_R + \varepsilon_2)} \right)^2 \right], \tag{4.37}$$

$$\beta_I = \frac{k_0}{2} \sqrt{\frac{\varepsilon_R \varepsilon_2}{\varepsilon_R + \varepsilon_2}} \frac{\varepsilon_I \varepsilon_2}{\varepsilon_R(\varepsilon_R + \varepsilon_2)}, \tag{4.38}$$

$$p_{1R} = -\frac{k_0 \varepsilon_R}{\sqrt{-\varepsilon_R - \varepsilon_2}} \left[1 + \frac{1}{8} \left(1 + 4\frac{\varepsilon_2}{\varepsilon_R} \right) \left(\frac{\varepsilon_I}{\varepsilon_R + \varepsilon_2} \right)^2 \right], \tag{4.39}$$

$$p_{1I} = \frac{k_0}{2} \frac{(\varepsilon_R + 2\varepsilon_2)\varepsilon_I}{\sqrt{-\varepsilon_R - \varepsilon_2}^3}, \tag{4.40}$$

$$p_{2R} = \frac{k_0 \varepsilon_2}{\sqrt{-\varepsilon_R - \varepsilon_2}} \left[1 - \frac{3}{8} \left(\frac{\varepsilon_I}{\varepsilon_R + \varepsilon_2} \right)^2 \right], \tag{4.41}$$

$$p_{2I} = \frac{k_0}{2} \frac{\varepsilon_I \varepsilon_2}{\sqrt{-\varepsilon_R - \varepsilon_2}^3}, \tag{4.42}$$

where we have assumed that $\varepsilon_R < -\varepsilon_2 < 0$. The imaginary parts of the three wave vectors are thus directly proportional to ε_I, which only participates as a second-order correction on the real parts of the wave vectors. Given that the fields exhibit and exponential dependence on the transverse direction ($\sim \exp(-p_n |x|)$), this proportionality ($p_{nI} \propto \varepsilon_I$) highlights the evanescent (nonpropagative) nature of polaritons along the cross-plane direction, for low absorption media ($\varepsilon_I \to 0$). For $\varepsilon_R > 0$, on the other hand, (4.37) and (4.38) for the in-plane wave vector are still valid, while (4.39)–(4.42) for the cross-plane ones have to be replaced with

$$p_{1R} = \frac{k_0}{2} \frac{(\varepsilon_R + 2\varepsilon_2)\varepsilon_I}{\sqrt{\varepsilon_R + \varepsilon_2}^3}, \tag{4.43}$$

$$p_{1I} = -\frac{k_0 \varepsilon_R}{\sqrt{\varepsilon_R + \varepsilon_2}} \left[1 + \frac{1}{8} \left(1 + 4\frac{\varepsilon_2}{\varepsilon_R} \right) \left(\frac{\varepsilon_I}{\varepsilon_R + \varepsilon_2} \right)^2 \right], \tag{4.44}$$

$$p_{2R} = \frac{k_0}{2} \frac{\varepsilon_I \varepsilon_2}{\sqrt{\varepsilon_R + \varepsilon_2}^3}, \tag{4.45}$$

$$p_{2I} = \frac{k_0 \varepsilon_2}{\sqrt{\varepsilon_R + \varepsilon_2}} \left[1 - \frac{3}{8} \left(\frac{\varepsilon_I}{\varepsilon_R + \varepsilon_2} \right)^2 \right]. \tag{4.46}$$

In contrast to the case of $\varepsilon_R < -\varepsilon_2 < 0$, for $\varepsilon_R > 0$, the real parts of the cross-plane wave vectors reduce linearly with ε_I, as shown by (4.43) and (4.45). Therefore, the existence and propagation of polaritons in polar materials tend to fade away for sufficiently low and high frequencies, for which $\varepsilon_I \to 0$, while ε_R takes positive values, as established by the Lorentz model in (3.36) and (3.37). This behavior also holds for the intermediate case of $-\varepsilon_2 < \varepsilon_R < 0$, as (4.43)–(4.46) keep their validity under these latter conditions. However, for $-\varepsilon_2 < \varepsilon_R < 0$, the real and imaginary parts of the in-plane wave vector are no longer given by (4.37) and (4.38), but rather by

$$\beta_R = \frac{k_0}{2}\sqrt{\frac{|\varepsilon_R|\,\varepsilon_2}{\varepsilon_R+\varepsilon_2}}\frac{\varepsilon_I\varepsilon_2}{|\varepsilon_R|\,(\varepsilon_R+\varepsilon_2)}, \tag{4.47}$$

$$\beta_I = k_0\sqrt{\frac{|\varepsilon_R|\,\varepsilon_2}{\varepsilon_R+\varepsilon_2}}\left[1+\frac{\varepsilon_2}{8}\,(\varepsilon_2+4\varepsilon_R)\left(\frac{\varepsilon_I}{\varepsilon_R(\varepsilon_R+\varepsilon_2)}\right)^2\right]. \tag{4.48}$$

As in the case of the cross-plane wave vectors, β_R decreases linearly with ε_I, while β_I is driven by ε_R mainly.

The real ε_R and imaginary ε_I parts of the relative permittivity $\varepsilon = \varepsilon_R + i\varepsilon_I$ of silicon carbide (SiC) are well determined by the Lorenz model in (3.35), and their frequency spectra are shown in Fig. 4.3. The main resonance peak of ε_I at $\omega \approx \omega_T = 149$ Trad/s (23.71 THz) indicates that SiC absorbs significant energy from the electromagnetic fields at that frequency. By contrast, the small values of ε_I outside of the vicinity of this peak show that these fields can propagate relatively long distances for frequencies smaller and greater than ω_T. The peak of ε_I also determines the lower bound of the yellow zone ($\omega_T \lesssim \omega \lesssim \omega_L$), where ε_R is negative (Reststrahlen band). This well-defined interval of frequencies is thus expected to favor the appearance and propagation of SPhPs along the interface of SiC.

The real part of the wave vector $\beta_R = \mathrm{Re}(\beta)$ and the wavelength $\lambda = 2\pi/\beta_R$ driving the propagation of polaritons along a single interface between SiC and vacuum (or air) are shown in Fig. 4.4, as functions of frequency. For $\omega < \omega_T$, β_R is practically superimposed with the light line ($\beta_R = k_0 = \omega/c$), which reveals the photon-like behavior of the electromagnetic fields. As the frequency

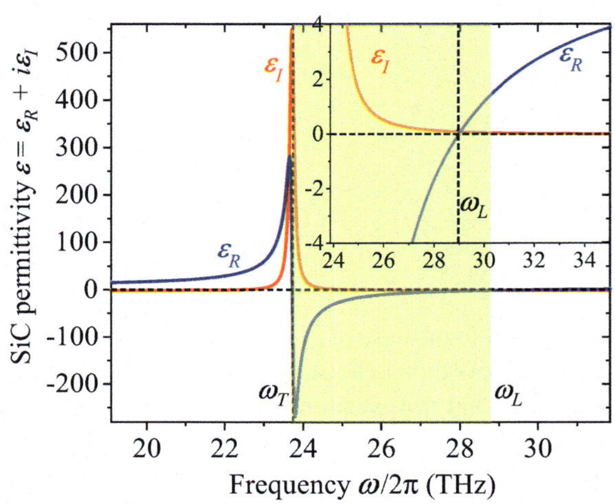

Fig. 4.3 Frequency spectrum of the real and imaginary parts of the relative permittivity of SiC. Calculations were done with the Lorentz model in (3.35) and the input data reported in Table 3.1. The real part is negative ($\varepsilon_R < 0$) inside the yellow zone, which is approximately determined by the interval $\omega_T \lesssim \omega \lesssim \omega_L$

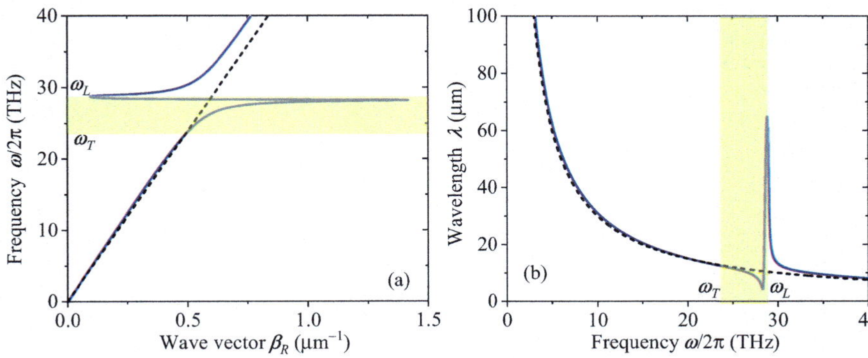

Fig. 4.4 Frequency spectrum of the (**a**) wave vector β_R and (**b**) wavelength $\lambda = 2\pi/\beta_R$ of polariton propagation along a single interface between SiC and vacuum. The dashed lines in (**a**) and (**b**) represent the wave vector k_0 and wavelength $2\pi/k_0$ in vacuum ($\varepsilon_2 = 1$), respectively

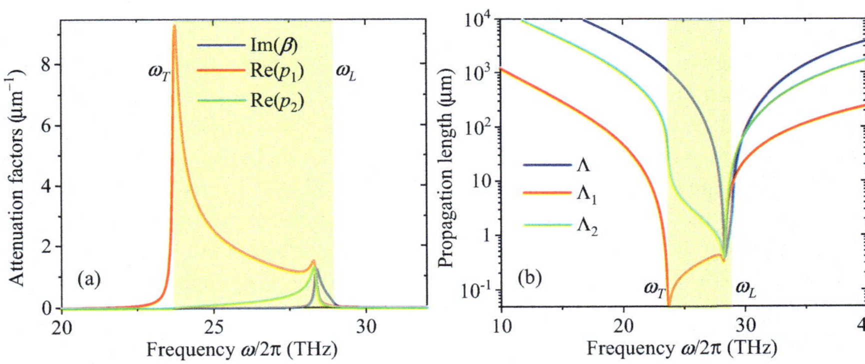

Fig. 4.5 Frequency spectrum of the in-plane and cross-plane (**a**) attenuation factors and (**b**) propagation lengths of polariton propagation along a single interface between SiC and vacuum

increases, the wave vector separates from the light line through higher values ($\beta_R > k_0$) in practically the entire yellow Reststrahlen band. High wave vectors (small wavelengths) therefore favor the appearance and propagation of polaritons. The fact that $\lambda > 1$ μm indicates that this propagation is not affected by the typical peaks of rugosity ($\ll 1$ μm) that surfaces usually exhibit.

The frequency spectrum of the factors exponentially attenuating the electromagnetic fields along the in-plane (Im(β)) and cross-plane (Re(p_n) for $n = 1$ and 2) directions of the interface SiC/vacuum is shown in Fig. 4.5a. Note that the cross-plane attenuation factor inside SiC is generally greater than the corresponding one in vacuum and along the SiC/vacuum interface (Re(p_1) $\gg$ Re(p_2) and Im(β)), due to the energy absorption of SiC within the Reststrahlen band. Outside of this yellow band, the cross-plane attenuation is negligible, and hence the electromagnetic field is

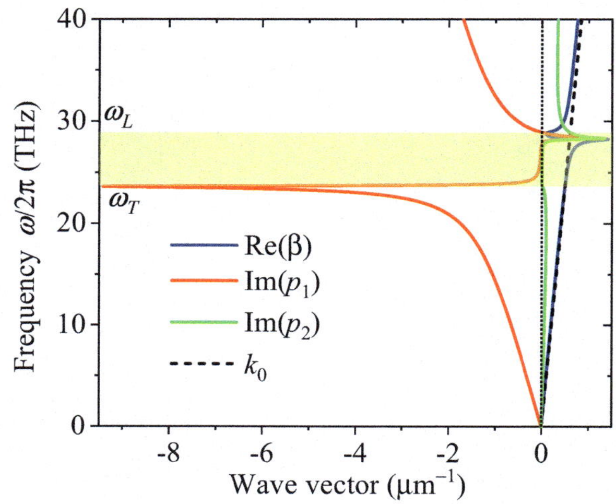

Fig. 4.6 Frequency spectrum of the in-plane and cross-plane wave vectors of polariton propagation along a single interface between SiC and vacuum

weakly confined to the propagation interface. The robust appearance of SPhPs inside the Reststrahlen band is thus accompanied with the shortest propagation distances along the in-plane ($\Lambda = 1/2\text{Im}(\beta)$) and cross-plane directions ($\Lambda_n = 1/2\text{Re}(p_n)$), as shown in Fig. 4.5b. Furthermore, the fact that the cross-plane wave vectors inside SiC ($\text{Im}(p_1)$) and vacuum ($\text{Im}(p_2)$) are nearly zero (Fig. 4.6), for the vast majority of frequencies inside the Reststrahlen band, reveals the nonpropagative nature of polaritons along the cross-plane direction. Polaritons are thus propagative ($\beta_R > 0$) in the in-plane direction and evanescent ($\Lambda_n > 0$) in the cross-plane one, as depicted in Fig. 4.2.

The combination of (4.19) and (4.20) with the expressions of the wave vectors in (4.22)–(4.24) allows expressing the field components for the TM polarization in terms of the material permittivities, as follows:

$$E_z^{(1)} = Ae^{p_1 x}, \quad E_x^{(1)} = i\sqrt{\frac{-\varepsilon_2}{\varepsilon_1}}E_z^{(1)}, \quad H_y^{(1)} = ic\varepsilon_0\sqrt{-\varepsilon_1 - \varepsilon_2}E_z^{(1)} \tag{4.49}$$

$$E_z^{(2)} = Ae^{-p_2 x}, \quad E_x^{(2)} = i\sqrt{\frac{-\varepsilon_1}{\varepsilon_2}}E_z^{(2)}, \quad H_y^{(2)} = ic\varepsilon_0\sqrt{-\varepsilon_1 - \varepsilon_2}E_z^{(2)}. \tag{4.50}$$

The spectral thermal energy generated by these fields inside the energy absorbing medium 1 can now be found by inserting (4.49) and (4.50) into (4.5). The final result is

$$\frac{\overline{Q}_1}{|A|^2} = \frac{\omega\varepsilon_0}{2}\text{Im}\left(\varepsilon_1\right)\left(1 + \frac{\varepsilon_2}{|\varepsilon_1|}\right)e^{x/\Lambda_1 - z/\Lambda}, \tag{4.51}$$

where $\Lambda = [2\text{Im}(\beta)]^{-1}$and $\Lambda_1 = [2\text{Re}(p_1)]^{-1}$ are the in-plane and cross-plane propagation lengths of the fields. Equation (4.51) thus explicitly shows that the generation of heat inside the energy absorbing material 1 is directly proportional to the imaginary part of its permittivity. No thermal energy is generated within the nonabsorbing material 2 ($\overline{Q}_2 = 0$) because Im $(\varepsilon_2) = 0$. However, this is not the case of the Poynting vector, which has nonvanishing in-plane and cross-plane components in both media. With the help of the fields in (4.49) and (4.50), we can rewrite (4.6) and (4.7) for each medium, as follows:

$$\frac{S_z^{(n)}}{|A|^2} = \frac{\omega\varepsilon_0}{2|p_n|^2}\text{Re}\left(\beta\varepsilon_n^*\right)e^{-|x|/\Lambda_n - z/\Lambda}, \tag{4.52}$$

$$\frac{S_x^{(n)}}{|A|^2} = \frac{c\varepsilon_0}{2\sqrt{2}}\sqrt{|\varepsilon_1 + \varepsilon_2| + \varepsilon_R + \varepsilon_2}e^{-|x|/\Lambda_n - z/\Lambda}, \tag{4.53}$$

where $\Lambda_n = [2\text{Re}(p_n)]^{-1}$ is the cross-plane propagation length of the fields in media $n = 1$ and 2. The exponential decay of both components of the Poynting vector is thus equal inside the medium n. Taking into account that $\Lambda_1 < \Lambda_2$, the energy density is highly confined inside and close to the interface of the absorbing medium 1, specially for frequencies within the Reststrahlen band. In the absence of energy absorption ($\text{Im}(\varepsilon_I) = 0$), $S_z^{(n)}$ and $S_x^{(n)}$ do not vanish, as the propagation of the electromagnetic (nonthermal) energy density is not limited by the generation of heat.

The typical spatial distribution of the magnetic field H of the TM waves propagating along an interface of SiC with vacuum is shown in Fig. 4.7. Note that H propagates hundreds of μm practically without attenuation along the in-plane direction, while it does exhibit a strong attenuation along the cross-plane direction. The distances that H travels inside vacuum ($x > 0$) and SiC ($x < 0$) are much longer and smaller than 1 μm, respectively. This surface confinement of the magnetic field inside the absorbing medium reveals that the TM waves are actually SPhPs, since the components of the electrical field have the same spatial dependence as the magnetic one, as established by (4.49) and (4.50). The propagation of the electromagnetic field inside and close to the surface of SiC thus

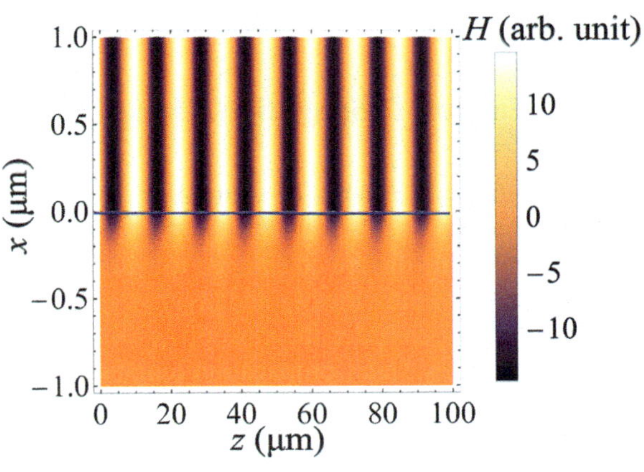

Fig. 4.7 Spatial distribution of the magnetic field $H = \text{Re}\left(H_y\exp(i\beta z)\right)$ of the TM waves propagating along an interface of SiC with vacuum. Calculations were done with (4.49) and (4.50), for the frequency $\omega = 150$ Trad/s inside the Reststrahlen band of the SiC permittivity

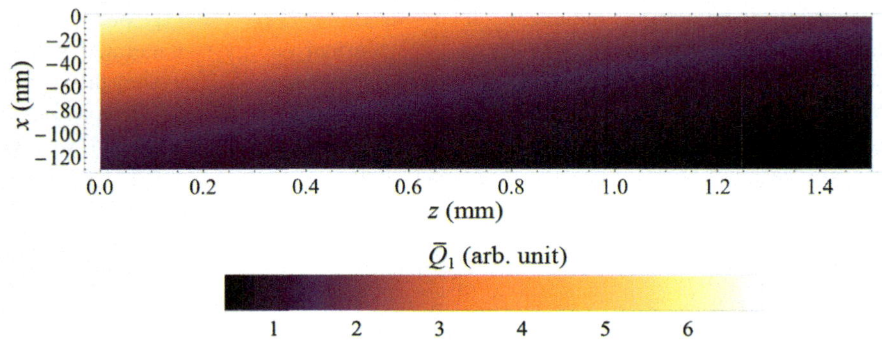

Fig. 4.8 Spatial distribution of the heat source $\overline{Q}_1$ generated by the TM waves propagating along the interface of SiC with vacuum. Calculations were done with (4.51) and for the frequency $\omega = 150$ Trad/s inside the Reststrahlen band of the SiC permittivity

Fig. 4.9 Propagation of polaritons along the interfaces $x = \pm d/2$ of a film of thickness d and permittivity ε_2

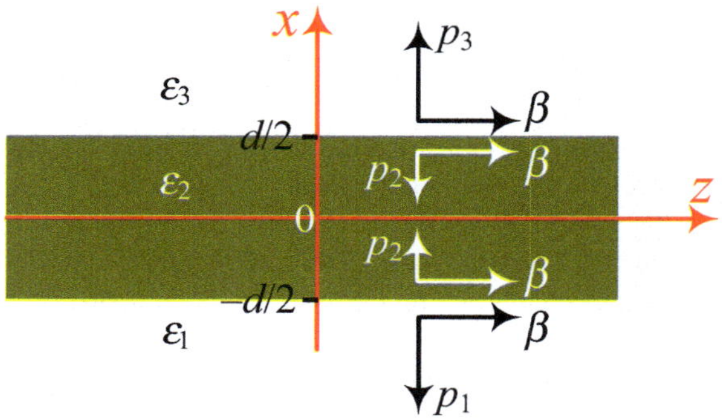

generates the thermal energy $\overline{Q}_1$ shown in Fig. 4.8. In the across-plane direction, $\overline{Q}_1$ is present over 100 nm only, while along the in-plane one, it shows up over 1 mm. This relatively long distance reveals the great potential of SPhPs as effective heat carriers.

4.3 Polariton Propagation Along Two Interfaces

This configuration consists of a thin film supporting the propagation of polaritons along its interfaces with two semi-infinite media, as shown in Fig. 4.9. The film and its surrounding media are considered to be nonmagnetic (magnetic permeability $= \mu_0$) and have permittivities ε_2 and (ε_1; ε_3), respectively. For the sake of simplicity, we are going to consider the case of $\varepsilon_3 = \varepsilon_1$, as the case of $\varepsilon_3 \neq \varepsilon_1$ generally leads to relatively small propagation lengths, as discussed in the literature [1–3]. The cross-plane wave vector p_n inside the media $n = 1$ and 2 is thus related to the in-plane one via (4.4), as follows: $p_n^2 = \beta^2 - \varepsilon_n k_0^2$. The frequency dependence of these wave vectors is given by the dispersion relation, which we are now going to derive for both the TM and TE polarizations.

4.3.1 *TM Polarization*

Considering that the electric and magnetic fields must vanish or remind finite for $x \to \pm\infty$, the in-plane components of the TM waves predicted by (4.2) and (4.3) can be written as follows:

$$E_z^{(1<)} = A^{<} e^{p_1(x+d/2)}, \quad H_y^{(1<)} = -\frac{i\omega\varepsilon_0\varepsilon_1}{p_1} E_z^{(1<)}, \tag{4.54}$$

$$E_z^{(1>)} = A^{>} e^{-p_1(x-d/2)}, \quad H_y^{(1>)} = \frac{i\omega\varepsilon_0\varepsilon_1}{p_1} E_z^{(1>)}, \tag{4.55}$$

$$E_z^{(2)} = A\sinh(p_2 x) + B\cosh(p_2 x), \quad H_y^{(2)} = -\frac{i\omega\varepsilon_0\varepsilon_2}{p_2^2}\frac{\partial E_z^{(2)}}{\partial x}, \tag{4.56}$$

where the superscripts (1 $<$) and (1 $>$) stand for medium 1 below ($x < -d/2$) and above ($x > d/2$) the thin film. The continuity of these tangential components at the boundary surfaces $x = \pm d/2$ yields

$$A^{<} = -A\sinh(a) + B\cosh(a) = S_{21}\left[A\cosh(a) - B\sinh(a)\right], \tag{4.57}$$

$$A^{>} = A\sinh(a) + B\cosh(a) = -S_{21}\left[A\cosh(a) + B\sinh(a)\right], \tag{4.58}$$

where $a = p_2 d/2$, $S_{21} = S_2/S_1$, and $S_n = \varepsilon_n/p_n$. The sum and difference of (4.57) and (4.58) yield

$$A^{<} + A^{>} = 2B\cosh(a) = -2BS_{21}\sinh(a), \tag{4.59}$$

$$A^{<} - A^{>} = -2A\sinh(a) = 2AS_{21}\cosh(a). \tag{4.60}$$

Equations (4.59) and (4.60) define two dispersion relations, which indicates that the propagation of the TM waves is driven by two modes.

TM Symmetric (Even) Mode

This mode is defined by the dispersion relation obtained for $A \neq 0$, which according to (4.60) yields

$$\frac{p_1}{\varepsilon_1} + \frac{p_2}{\varepsilon_2}\tanh\left(\frac{p_2 d}{2}\right) = 0. \tag{4.61}$$

Equation (4.61) in combination with (4.59) and (4.60) establishes that $B = 0$ and $A^{>} = -A^{<} = A\sinh(a)$, and therefore the magnetic field inside the film is an even $\left(H_y^{(2)}(-x) = H_y^{(2)}(x)\right)$ function of position x. This symmetric behavior of H_y provides the name of the even mode of the TM waves. The dispersion relation (4.61) is a transcendental equation that cannot be solved analytically for the wave vector β and an arbitrary film thickness d. However, for a very thick film ($|p_2|d \gg 2$), the hyperbolic tangent $\tanh(p_2 d/2) \to 1$ and (4.61) becomes independent of d and reduces to the dispersion relation (4.21) for a single interface, as expected. On the other hand, for a very thin film ($|p_2|d \ll 2$), (4.61) can approximately be solved by rewriting it in terms of the effective permittivity $\varepsilon = (\beta/k_0)^2$, as follows:

$$\varepsilon_2\sqrt{\varepsilon-\varepsilon_1} + \varepsilon_1\sqrt{\varepsilon-\varepsilon_2}\tanh\left(h\sqrt{\varepsilon-\varepsilon_2}\right) = 0, \tag{4.62}$$

where $h = k_0 d/2$, and we have used (4.4) to write the cross-plane wave vectors in the form $p_n = k_0\sqrt{\varepsilon - \varepsilon_n}$. For a very thin film ($h \ll 1$) and an approximation up to h^4, the Taylor series expansion of the hyperbolic tangent $\tanh\left(h\sqrt{\varepsilon-\varepsilon_2}\right) \approx h\sqrt{\varepsilon-\varepsilon_2}\left[1 - 3^{-1}h^2(\varepsilon-\varepsilon_2)\right]$ transforms (4.62) into the following polynomial equation:

$$\varepsilon_2\sqrt{\varepsilon-\varepsilon_1} + \varepsilon_1 h(\varepsilon-\varepsilon_2)\left[1 - \frac{h^2}{3}(\varepsilon-\varepsilon_2)\right] = 0. \tag{4.63}$$

The dependence of (4.63) on h indicates that its solution for ε is given by the following expansion:

$$\varepsilon = \varepsilon_1 + (\alpha h)^2\left(1 + a_2 h^2\right), \tag{4.64}$$

where the parameters α and a_2 are determined by inserting (4.64) into (4.63). The final result is

$$\alpha = \varepsilon_1\left(1 - \frac{\varepsilon_1}{\varepsilon_2}\right), \tag{4.65}$$

$$a_2 = \frac{2}{3}(\varepsilon_2 - \varepsilon_1)\left[1 - 3\left(\frac{\varepsilon_1}{\varepsilon_2}\right)^2\right]. \tag{4.66}$$

Equations (4.64)–(4.66) thus establish that the wave vector $\beta = k_0\sqrt{\varepsilon}$ is driven by the ratio $\varepsilon_1/\varepsilon_2$ and difference $\varepsilon_2 - \varepsilon_1$ of permittivities. For an approximation up to h^3, one obtains

$$\beta = k_0\sqrt{\varepsilon_1}\left[1 + \frac{\varepsilon_1}{2}\left(1 - \frac{\varepsilon_1}{\varepsilon_2}\right)^2 h^2\right]. \tag{4.67}$$

Considering that the film with permittivity $\varepsilon_2 = \varepsilon_R + i\varepsilon_I$ is surrounded by a nonabsorbing medium ($\varepsilon_1 > 0$), (4.67) yields the following real and imaginary parts of $\beta = \beta_R + i\beta_I$:

$$\beta_R = k_0\sqrt{\varepsilon_1}\left\{1 + \frac{\varepsilon_1 h^2}{2}\left[1 - 2\frac{\varepsilon_1\varepsilon_R}{|\varepsilon_2|^2} + \frac{\varepsilon_1^2\left(\varepsilon_R^2 - \varepsilon_I^2\right)}{|\varepsilon_2|^4}\right]\right\}, \tag{4.68}$$

$$\beta_I = k_0\varepsilon_1^{5/2}\frac{\varepsilon_I h^2}{|\varepsilon_2|^2}\left(1 - \frac{\varepsilon_1\varepsilon_R}{|\varepsilon_2|^2}\right), \tag{4.69}$$

where $|\varepsilon_2|^2 = \varepsilon_R^2 + \varepsilon_I^2$. Note that the square of the normalized film thickness $h = k_0 d/2$ is the second leading term of β_R, but the dominant one for β_I. Therefore, as the film thickness reduces, the TM waves propagate with a wave vector $\beta_R \to k_0\sqrt{\varepsilon_1}$ nearly independent of the film properties, while their propagation length $\Lambda = (2\beta_I)^{-2} \propto h^{-2}$ increases without limit, as is the case of nonabsorbing films ($\varepsilon_I = 0$). This increase thus shows that the even mode of the TM waves is a long-range mode, when propagating along very thin films. Furthermore, (4.69) shows that the symmetric mode propagates ($\Lambda > 0$) provided that $\varepsilon_R^2 + \varepsilon_I^2 - \varepsilon_1\varepsilon_R > 0$ [4]. This constraint establishes that the propagation of these TM waves appears for all permittivity components ε_R and ε_I outside the semicircle centered at $(\varepsilon_1/2; 0)$ with a radius $\varepsilon_1/2$, as shown in Fig. 4.10. This propagation condition also ensures the existence of polaritons, as it guarantees the surface confinement ($p_{1R} > 0$) of the symmetric mode propagating outside the film. This can be seen by inserting the effective permittivity in (4.64) into the cross-plane wave vector $p_1 = p_{1R} + ip_{1I} = k_0\sqrt{\varepsilon - \varepsilon_1}$, which yields

$$p_{1R} = k_0\varepsilon_1\left(1 - \frac{\varepsilon_1\varepsilon_R}{|\varepsilon_2|^2}\right)h, \quad p_{1I} = k_0\frac{\varepsilon_1^2\varepsilon_I}{|\varepsilon_2|^2}h. \tag{4.70}$$

The surface confinement ($p_{1R} \propto h$) thus linearly reduces with the film thickness. The cross-plane wave vector $p_2 = k_0\sqrt{\varepsilon - \varepsilon_2}$ inside the film, on the other hand, has only a second-order dependence on the film thickness and is given by

$$p_2 = k_0\sqrt{\varepsilon_1 - \varepsilon_2}\left[1 + \frac{(\varepsilon_1 h)^2}{2}(\varepsilon_1 - \varepsilon_2)\right]. \tag{4.71}$$

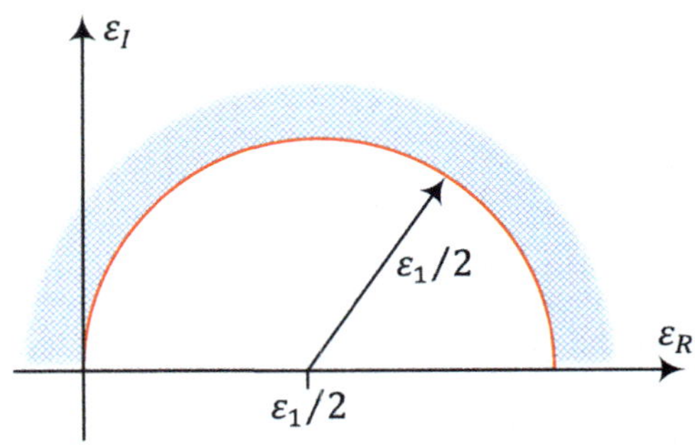

Fig. 4.10 Bluish zone in the permittivity plane ($\varepsilon_R; \varepsilon_I$), supporting the propagation of the symmetric TM modes along a thin film

The dispersion relation in (4.61) along with the wave vectors in (4.68)–(4.71) determines the electromagnetic fields of the TM symmetric mode. As the mode is driven by $A^{>} = -A^{<} = A\sinh(p_2 d/2)$, the tangential components in (4.54)–(4.56) and the orthogonal one in (4.1) can be written as follows:

$$E_z^{(1)} = A\text{sign}(x)\sinh(p_2 d/2)\, e^{-p_1(|x|-d/2)}, \tag{4.72}$$

$$E_x^{(1)} = \frac{i\beta}{p_1} A\sinh(p_2 d/2)\, e^{-p_1(|x|-d/2)}, \tag{4.73}$$

$$H_y^{(1)} = \frac{i\omega\varepsilon_0\varepsilon_1}{p_1} A\sinh(p_2 d/2)\, e^{-p_1(|x|-d/2)}, \tag{4.74}$$

$$E_z^{(2)} = A\sinh(p_2 x), \tag{4.75}$$

$$E_x^{(2)} = -\frac{i\beta}{p_2} A\cosh(p_2 x), \tag{4.76}$$

$$H_y^{(2)} = \frac{i\omega\varepsilon_0\varepsilon_2}{p_2} A\cosh(p_2 x), \tag{4.77}$$

where $|x| = x$ and $\text{sign}(x) = 1$, for $x > d/2$, and $|x| = -x$ and $\text{sign}(x) = -1$, for $x < -d/2$. The combination of (4.75) and (4.76) with (4.5) establishes that the spatial distribution of the spectral thermal energy generated by these fields inside the thin film is given by

$$\frac{\overline{Q}_2}{|A|^2} = \frac{\omega\varepsilon_0}{2}\text{Im}(\varepsilon_2)\left[|\sinh(p_2 x)|^2 + \left|\frac{\beta}{p_2}\right|^2 |\cosh(p_2 x)|^2\right] e^{-z/\Lambda}. \tag{4.78}$$

Given that $|\sinh(u+iv)|^2 = \sinh(u+iv)\sinh(u-iv) = 0.5\,[\cosh(2u) - \cos(2v)]$ and $|\cosh(u+iv)|^2 = \cosh(u+iv)\cosh(u-iv) = 0.5\,[\cosh(2u) + \cos(2v)]$, for any complex number $u + iv$, the dependence of $\overline{Q}_2$ on the cross-plane position x can be written in terms of the cross-plane propagation length $\Lambda_2 = [2\text{Re}(p_2)]^{-1}$ and cross-plane wavelength $\lambda_2 = 2\pi/\text{Im}(p_2)$, as follows:

$$\frac{\overline{Q}_2}{|A|^2} = \frac{\omega\varepsilon_0}{4}\text{Im}(\varepsilon_2)\left[\left(\left|\frac{\beta}{p_2}\right|^2 + 1\right)\cosh\left(\frac{x}{\Lambda_2}\right) + \left(\left|\frac{\beta}{p_2}\right|^2 - 1\right)\cos\left(\frac{4\pi x}{\lambda_2}\right)\right] e^{-z/\Lambda}. \tag{4.79}$$

The monotonic increase of the hyperbolic cosine with its argument thus indicates that the maximum heat generated by the symmetric TM waves inside the thin film appears at its interfaces $x = \pm d/2$, as expected for polaritons. The minimum of $\overline{Q}_2$, on the other hand, shows up at the middle ($x = 0$) of the film. This is also the case of the in-plane Poynting vector, which has nonvanishing in-plane and cross-plane components inside and outside the film. After inserting (4.72)–(4.77) into (4.6) and (4.7), one obtains

$$\frac{S_z^{(1)}}{|A|^2} = \frac{\omega\varepsilon_0}{2\,|p_1|^2}\mathrm{Re}\left(\beta\varepsilon_1^*\right)|\sinh(p_2 d/2)|^2\, e^{-(|x|-d/2)/\Lambda_1 - z/\Lambda}, \tag{4.80}$$

$$\frac{S_z^{(2)}}{|A|^2} = \frac{\omega\varepsilon_0}{4\,|p_2|^2}\mathrm{Re}\left(\beta\varepsilon_2^*\right)\left[\cosh\left(\frac{x}{\Lambda_2}\right) + \cos\left(\frac{4\pi x}{\lambda_2}\right)\right] e^{-z/\Lambda}, \tag{4.81}$$

$$\frac{S_x^{(1)}}{|A|^2} = \frac{\omega\varepsilon_0}{2}\mathrm{sign}(x)\,|\sinh(p_2 d/2)|^2\, \mathrm{Im}\left(\frac{\varepsilon_1}{p_1}\right) e^{-(|x|-d/2)/\Lambda_1 - z/\Lambda}, \tag{4.82}$$

$$\frac{S_x^{(2)}}{|A|^2} = -\frac{\omega\varepsilon_0}{4}\mathrm{Im}\left[\frac{\varepsilon_2}{p_2}\left(\sinh\left(\frac{x}{\Lambda_2}\right) - i\sin\left(\frac{4\pi x}{\lambda_2}\right)\right)\right] e^{-z/\Lambda}, \tag{4.83}$$

where we have used the identity $\sinh(u - iv)\cosh(u + iv) = 0.5(\sinh(2u) - i\sin(2v))$, for the number $u + iv$ and its complex conjugate $u - iv$. The in-plane and cross-plane components of the Poynting vector outside the film thus take its maxima at the film interfaces $x = \pm d/2$ and decay exponentially as the electromagnetic energy density travels away from them. By contrast, inside the film, this decay is not purely exponential, but the energy density still takes its maximum and minimum values at the interfaces and center ($x = 0$) of the film, respectively. Therefore, the thermal and electromagnetic energy densities of the symmetric TM waves take their maxima and minima at the same positions, as established by (4.79), (4.81), and (4.83).

TM Antisymmetric (Odd) Mode

The dispersion relation of this mode is determined by the condition $B \neq 0$ in (4.59), which yields

$$\frac{p_2}{\varepsilon_2} + \frac{p_1}{\varepsilon_1}\tanh\left(\frac{p_2 d}{2}\right) = 0. \tag{4.84}$$

The combination of (4.84) with (4.59) and (4.60) then establishes that $A = 0$ and $A^> = A^< = B\cosh(a)$, which transforms the magnetic field inside the film is an odd $\left(H_y^{(2)}(-x) = -H_y^{(2)}(x)\right)$ function of position x. This antisymmetric behavior of H_y provides the name of the odd mode of the TM waves. As in the case of the even mode, the dispersion relation (4.84) becomes independent of d and reduces to the one for a single interface, for a very thick film ($|p_2|\,d \gg 2$). On the other hand, for a very thin film ($|p_2|\,d \ll 2$), the approximate solution of (4.84) for $\beta = k_0\sqrt{\varepsilon}$ can be obtained by rewriting it in terms of the effective permittivity ε, as follows:

$$\varepsilon_1\sqrt{\varepsilon - \varepsilon_2} + \varepsilon_2\sqrt{\varepsilon - \varepsilon_1}\tanh\left(h\sqrt{\varepsilon - \varepsilon_2}\right) = 0. \tag{4.85}$$

For a thin enough film ($h = k_0 d/2 \ll 1$) and an approximation up to h^4, the Taylor series expansion of $\tanh\left(h\sqrt{\varepsilon-\varepsilon_2}\right) \approx h\sqrt{\varepsilon-\varepsilon_2}\left[1-3^{-1}h^2\left(\varepsilon-\varepsilon_2\right)\right]$ transforms (4.85) into the following polynomial equation:

$$\varepsilon_1+\varepsilon_2 h\sqrt{\varepsilon-\varepsilon_1}\left[1-\frac{h^2}{3}\left(\varepsilon-\varepsilon_2\right)\right]=0. \tag{4.86}$$

The square root in (4.86) thus indicates that its solution for ε has the following mathematical form:

$$\varepsilon=\varepsilon_1+\left(\frac{\alpha}{h}\right)^2\left(1+a_2 h^2+a_4 h^4\right), \tag{4.87}$$

where the parameters α, a_2, and a_4 are determined by inserting (4.87) into (4.86). The final result is

$$\alpha^3-3\alpha-3\frac{\varepsilon_1}{\varepsilon_2}=0, \tag{4.88}$$

$$a_2=\frac{2}{3}\frac{\alpha\left(\varepsilon_2-\varepsilon_1\right)}{2\alpha+3\varepsilon_1/\varepsilon_2}, \tag{4.89}$$

$$a_4=\frac{\varepsilon_1}{\varepsilon_2}\frac{\left(\alpha+\varepsilon_1/\varepsilon_2\right)\left(\varepsilon_2-\varepsilon_1\right)^2}{\alpha\left(2\alpha+3\varepsilon_1/\varepsilon_2\right)^3}. \tag{4.90}$$

The three solutions of the depressed cubic equation (4.88) for α are given by the Cardano formula, as follows:

$$\alpha=\sqrt[3]{\frac{3\varepsilon_1}{2\varepsilon_2}+\sqrt{\Delta}}\,e^{2\pi m i/3}+\sqrt[3]{\frac{3\varepsilon_1}{2\varepsilon_2}-\sqrt{\Delta}}\,e^{-2\pi m i/3}, \tag{4.91}$$

where $m = 0, 1, 2$. Among these three mathematical solutions, the physical one is given by that for which $\mathrm{Im}(\varepsilon) > 0$. Since $\beta = k_0\sqrt{\varepsilon}$, this condition ensures the wave propagation ($\Lambda = [2\mathrm{Im}(\beta)]^{-1} > 0$). Therefore, the physical solution can be identified by plotting the imaginary part of the three solutions of ε, once the material permittivities ε_1 and ε_2 are defined. The in-plane $\beta = k_0\sqrt{\varepsilon}$ and cross-plane wave vectors $p_n = k_0\sqrt{\varepsilon-\varepsilon_n}$ are thus determined by the effective permittivity in (4.87). For a sufficiently thin film and relatively small emissivities ε_1 and ε_2, these wave vectors are equal and given by $\beta \approx p_1 \approx_2 \approx k_0\alpha/h$. The in-plane propagation length $\Lambda = [2\mathrm{Im}(\beta)]^{-1} \propto h$ hence linearly decreases with the film thickness. This reduction thus shows that the odd mode of the TM waves is a short-range mode, when propagating along very thin films.

The electromagnetic fields in (4.54)–(4.56) for the TM antisymmetric mode can now be rewritten with the help of the relation $A^{>} = A^{<} = B\cosh\left(p_2 d/2\right)$, which yields

$$E_z^{(1)} = B\cosh\left(p_2 d/2\right) e^{-p_1(|x|-d/2)}, \tag{4.92}$$

$$E_x^{(1)} = \frac{i\beta}{p_1} B\mathrm{sign}(x)\cosh\left(p_2 d/2\right) e^{-p_1(|x|-d/2)}, \tag{4.93}$$

$$H_y^{(1)} = \frac{i\omega\varepsilon_0\varepsilon_1}{p_1} B\mathrm{sign}(x)\cosh\left(p_2 d/2\right) e^{-p_1(|x|-d/2)}, \tag{4.94}$$

$$E_z^{(2)} = B\cosh\left(p_2 x\right), \tag{4.95}$$

$$E_x^{(2)} = -\frac{i\beta}{p_2} B\sinh\left(p_2 x\right), \tag{4.96}$$

$$H_y^{(2)} = -\frac{i\omega\varepsilon_0\varepsilon_2}{p_2} B\sinh\left(p_2 x\right). \tag{4.97}$$

The electric field components in (4.95) and (4.96) establish that the spatial distribution of the heat source in (4.5) generated by the odd mode of the TM waves inside the thin film is given by

$$\begin{aligned}\frac{\overline{Q}_2}{|B|^2} = \frac{\omega\varepsilon_0}{4}\mathrm{Im}\left(\varepsilon_2\right)\Bigg[&\left(1+\left|\frac{\beta}{p_2}\right|^2\right)\cosh\left(\frac{x}{\Lambda_2}\right)\\&+\left(1-\left|\frac{\beta}{p_2}\right|^2\right)\cos\left(\frac{4\pi x}{\lambda_2}\right)\Bigg]e^{-z/\Lambda}.\end{aligned} \tag{4.98}$$

The thermal energy of the odd mode thus exhibits a similar dependence on the cross-plane position to that in (4.79) for the even mode, through the ratio x/Λ_2 and x/λ_2. $\overline{Q}_2$ takes its highest values at the film interfaces $x = \pm d/2$ and its lowest ones at the film center $x = 0$. However, taking into account that $\beta \approx p_1 \approx_2 \approx k_0\alpha/h$ and $\Lambda_2 \propto d$, the thermal energy generated by the odd mode inside the film is expected to be nearly independent of the film thickness at the film interfaces, for a very thin film. The in-plane and cross-plane components of the Poynting vector, on the other hand, exhibit a similar spatial distribution to $\overline{Q}_2$. By inserting (4.92)–(4.97) into (4.6) and (4.7), one obtains

$$\frac{S_z^{(1)}}{|B|^2} = \frac{\omega\varepsilon_0}{2\left|p_1\right|^2}\mathrm{Re}\left(\beta\varepsilon_1^*\right)\left|\cosh(p_2 d/2)\right|^2 e^{-(|x|-d/2)/\Lambda_1 - z/\Lambda}, \tag{4.99}$$

$$\frac{S_z^{(2)}}{|B|^2} = \frac{\omega\varepsilon_0}{4\left|p_2\right|^2}\mathrm{Re}\left(\beta\varepsilon_2^*\right)\left[\cosh\left(\frac{x}{\Lambda_2}\right) - \cos\left(\frac{4\pi x}{\lambda_2}\right)\right]e^{-z/\Lambda}, \tag{4.100}$$

$$\frac{S_x^{(1)}}{|B|^2} = \frac{\omega\varepsilon_0}{2}\mathrm{sign}(x)\left|\cosh(p_2 d/2)\right|^2 \mathrm{Im}\left(\frac{\varepsilon_1}{p_1}\right) e^{-(|x|-d/2)/\Lambda_1 - z/\Lambda}, \tag{4.101}$$

$$\frac{S_x^{(2)}}{|B|^2} = -\frac{\omega\varepsilon_0}{4}\mathrm{Im}\left[\frac{\varepsilon_2}{p_2}\left(\sinh\left(\frac{x}{\Lambda_2}\right) + i\sin\left(\frac{4\pi x}{\lambda_2}\right)\right)\right]e^{-z/\Lambda}. \tag{4.102}$$

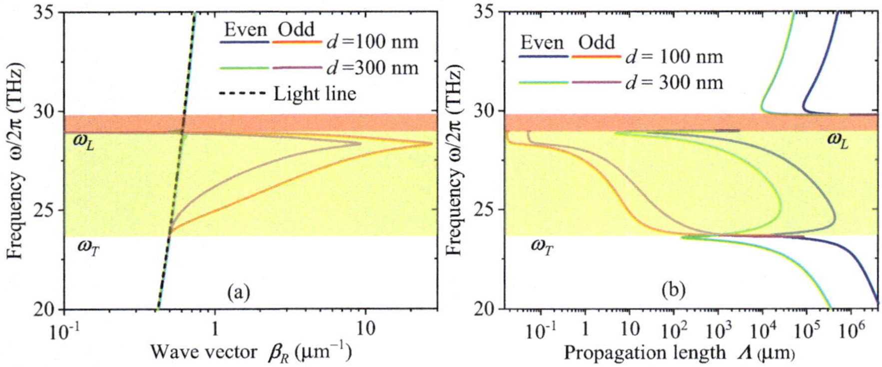

Fig. 4.11 Frequency spectrum of the in-plane (**a**) wave vector β_R and (**b**) propagation length of the TM waves propagating along the interfaces of a SiC nanofilm suspended in vacuum. Calculations were done for the even and odd modes and two film thicknesses. The yellow zone represents the Reststrahlen band shown in Fig. 4.3, while the red one is a forbidden band for which the dispersion relation of the even mode does not have a solution

The comparison of (4.99)–(4.102) with (4.80)–(4.83) shows that the electromagnetic energy density of the even and odd TM modes propagating inside and outside the film has pretty similar spatial distribution. However, their values are expected to be quite different, as they are driven by the dissimilar dispersion relations in (4.61) and (4.84).

The in-plane wave vector β_R and propagation length Λ of the even and odd modes of the TM waves propagating along the surfaces of a SiC nanofilm are, respectively, shown in Fig. 4.11a, b, as functions of frequency. Note that the even mode propagates with frequencies inside and outside of the Reststrahlen band ($\omega_T \lesssim \omega \lesssim \omega_L$), except for a relatively narrow forbidden band (red zone), in which the dispersion relation in (4.61) does not have a physical solution. While β_R(even mode) does not change significantly with the film thickness (for $d = 100$ and 300 nm), longer propagation lengths Λ(even mode) are obtained for thinner films, as predicted by (4.68) and (4.69). By contrast, the frequency spectrum of the odd mode is limited by the Reststrahlen band, such that it propagates with larger wave vectors and shorter propagation lengths than the corresponding ones of the even mode (β_R(odd mode) $>$ β_R(even mode) and Λ(odd mode) $<$ Λ(even mode)). Compared to the odd mode, the even mode is thus a long-range mode able to propagate in a broad spectrum of frequencies. Furthermore, Λ(odd mode) increases and β_R(odd mode) decreases, as the film thickens. Therefore, the propagation length of the even and odd modes of the TM waves exhibits an opposite dependence on the film thickness. The experimental proof of the existence of those TM waves was provided in our previous works where the dispersion relation and the propagation length spectrum of a silica film 500nm in thickness were measured [5].

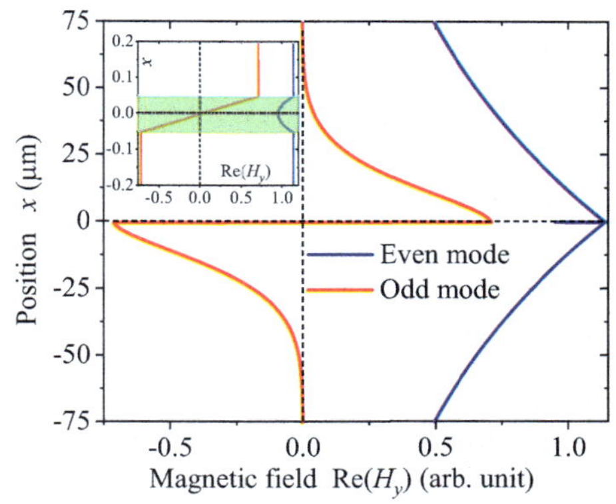

Fig. 4.12 Spatial distribution of the magnetic field Re(H_y) of the TM even and odd modes propagating along the interfaces of a 100-nm-thick film of SiC. Calculations were done for $\omega = 150$ Trad/s. The green zone in the inset stands for the SiC nanofilm

The magnetic field Re(H_y) of the TM even and odd modes propagating along a SiC nanofilm is shown in Fig. 4.12, as a function of the cross-plane position x. The symmetric and antisymmetric behavior of Re(H_y) around $x = 0$ defines the name of the even and odd modes, respectively. Note that the even mode can propagate hundreds of micrometers outside the nanofilm and exhibits a weaker confinement than the odd mode. This fact indicates that the even mode can significantly transport energy not only along the interfaces of the nanofilm but also outside it.

4.3.2 TE Polarization

The propagation of the TE waves can be analyzed by following similar steps as we developed for the TM ones. Taking into account the physical behavior (convergence) of the electric and magnetic fields for $x \to \pm\infty$, the in-plane components of the TE waves predicted by (4.9) and (4.10) can be written as follows:

$$H_z^{(1<)} = A^{<} e^{p_1(x+d/2)}, \quad E_y^{(1<)} = \frac{i\omega\mu_0}{p_1} H_z^{(1<)} \tag{4.103}$$

$$H_z^{(1>)} = A^{>} e^{-p_1(x-d/2)}, \quad E_y^{(1>)} = -\frac{i\omega\mu_0}{p_1} H_z^{(1>)} \tag{4.104}$$

$$H_z^{(2)} = A\sinh(p_2 x) + B\cosh(p_2 x), \quad E_y^{(2)} = -\frac{i\omega\mu_0}{p_2^2}\frac{\partial H_z^{(2)}}{\partial x}. \tag{4.105}$$

The continuity of these tangential components at the boundary surfaces $x = \pm d/2$ yields

$$A^{<} = -A\sinh(a) + B\cosh(a) = p_{12}\left[A\cosh(a) - B\sinh(a)\right], \tag{4.106}$$

$$A^{>} = A\sinh(a) + B\cosh(a) = -p_{12}\left[A\cosh(a) + B\sinh(a)\right], \tag{4.107}$$

where $a = p_2d/2$ and $p_{12} = p_1/p_2$. The sum and difference of (4.57) and (4.58) yield

$$A^< + A^> = 2B\cosh(a) = -2Bp_{12}\sinh(a), \tag{4.108}$$

$$A^< - A^> = -2A\sinh(a) = 2Ap_{12}\cosh(a). \tag{4.109}$$

As in the case of the TM waves, (4.108) and (4.109) show that the propagation of the TE waves is also driven by two modes.

TE Symmetric (Even) Mode
Considering that $A \neq 0$, the following dispersion relation is obtained from (4.109):

$$p_1 + p_2\tanh\left(\frac{p_2d}{2}\right) = 0. \tag{4.110}$$

The combination of (4.110) with (4.108) and (4.109) determines that $B = 0$ and $A^> = -A^< = A\sinh(a)$, which converts the electric field inside the film is an even $\left(E_y^{(2)}(-x) = E_y^{(2)}(x)\right)$ function of position x. This symmetry of H_y is the reason why this mode is called the even mode of the TE waves. The dispersion relation (4.111) is pretty similar to the one of the TM even mode in (4.60), and therefore its solution for the wave vector $\beta = k_0\sqrt{\varepsilon}$ can conveniently be obtained in terms of the effective permittivity ε given in (4.64), for a thin enough film ($h = k_0d/2 \ll 1$) and an approximation up to h^4. The parameters α and a_2 can be determined by inserting (4.110) into the Taylor series expansion of (4.64), which yields $\alpha = \varepsilon_2 - \varepsilon_1$ and $a_2 = -4\alpha/3$. The TE even mode is thus characterized by the following effective permittivity

$$\varepsilon = \varepsilon_1 + [(\varepsilon_2 - \varepsilon_1)h]^2\left[1 + \frac{4}{3}(\varepsilon_1 - \varepsilon_2)h^2\right]. \tag{4.111}$$

The real and imaginary parts of the in-plane wave vector $\beta = \beta_R + \beta_I = k_0\sqrt{\varepsilon}$ can readily be obtained from (4.111). For the sake of simplicity and brevity, their expressions for an approximation up to h^2 are

$$\beta_R = k_0\sqrt{\varepsilon_1}\left\{1 + \frac{h^2}{2\varepsilon_1}\left[(\varepsilon_R - \varepsilon_1)^2 - \varepsilon_I^2\right]\right\}, \tag{4.112}$$

$$\beta_I = k_0\frac{\varepsilon_I h^2}{\sqrt{\varepsilon_1}}(\varepsilon_R - \varepsilon_1). \tag{4.113}$$

The TE even mode thus propagates ($\Lambda = (2\beta_I)^{-1} > 0$) with a wave vector close to $\beta_R \to k_0\sqrt{\varepsilon_1}$, for all frequencies satisfying the condition $\varepsilon_R > \varepsilon_1$. For a thin film suspended in vacuum or air ($\varepsilon_1 = 1$), this condition indicates that the even mode of the TE waves propagates in a relatively limited range of frequencies, supporting low energy absorption ($\mathrm{Im}(\varepsilon_2) \to 0$), as shown in Fig. 4.3 for SiC. Therefore, the TE even mode is not expected to significantly contribute to the heat generation ($\overline{Q}_2 \propto \mathrm{Im}(\varepsilon_2)$) inside a thin film of SiC. The propagation condition ($\varepsilon_R > \varepsilon_1$) also ensures the surface confinement ($p_{1R} > 0$) of the mode outside the film, given that $p_n = k_0\sqrt{\varepsilon - \varepsilon_n}$, which yields

$$p_1 = k_0 h\left(\varepsilon_2 - \varepsilon_1\right), \tag{4.114}$$

$$p_2 = k_0\sqrt{\varepsilon_1 - \varepsilon_2}\left[1 - \frac{h^2}{2}\left(\varepsilon_1 - \varepsilon_2\right)\right]. \tag{4.115}$$

The outer surface confinement ($p_{1R} = k_0 h\left(\varepsilon_R - \varepsilon_1\right) > 0$) thus linearly reduces with the film thickness, while the cross-plane wave vector inside the film becomes independent of its thickness ($p_2 \to k_0\sqrt{\varepsilon_1 - \varepsilon_2}$), for thin enough films.

Taking into account that $B = 0$ and $A^{>} = -A^{<} = A\sinh(a)$, the field components of the TE even mode in (4.103)–(4.105) can now be written in terms of the single unknown A, as follows:

$$H_z^{(1)} = A\mathrm{sign}(x)\sinh\left(p_2 d/2\right)e^{-p_1(|x|-d/2)}, \tag{4.116}$$

$$H_x^{(1)} = \frac{i\beta}{p_1}A\sinh\left(p_2 d/2\right)e^{-p_1(|x|-d/2)}, \tag{4.117}$$

$$E_y^{(1)} = -\frac{i\omega\mu_0}{p_1}A\sinh\left(p_2 d/2\right)e^{-p_1(|x|-d/2)}, \tag{4.118}$$

$$H_z^{(2)} = A\sinh\left(p_2 x\right), \tag{4.119}$$

$$H_x^{(2)} = -\frac{i\beta}{p_2}A\cosh\left(p_2 x\right), \tag{4.120}$$

$$E_y^{(2)} = \frac{i\omega\mu_0}{p_2}A\cosh\left(p_2 x\right). \tag{4.121}$$

The tangential electric field component in (4.121) determines the spatial distribution of the thermal energy generated by the TE even mode inside the thin film, as established by (4.11). Inserting (4.121) into (4.11), one obtains

$$\frac{\overline{Q}_2}{|A|^2} = \frac{\omega\mu_0 k_0^2}{4\left|p_2\right|^2}\mathrm{Im}\left(\varepsilon_2\right)\left[\cosh\left(\frac{x}{\Lambda_2}\right) + \cos\left(\frac{4\pi x}{\lambda_2}\right)\right]e^{-z/\Lambda}. \tag{4.122}$$

Furthermore, the in-plane and cross-plane components of the Poynting vector inside and outside the thin film can be determined by inserting (4.116) and (4.119) into (4.12) and (4.13). This combination yields

$$\frac{S_z^{(1)}}{|A|^2} = \frac{\omega\mu_0}{2\,|p_1|^2}\text{Re}\,(\beta)\,|\sinh(p_2 d/2)|^2\, e^{-(|x|-d/2)/\Lambda_1 - z/\Lambda}, \quad (4.123)$$

$$\frac{S_z^{(2)}}{|A|^2} = \frac{\omega\mu_0}{4\,|p_2|^2}\text{Re}\,(\beta)\left[\cosh\left(\frac{x}{\Lambda_2}\right) + \cos\left(\frac{4\pi x}{\lambda_2}\right)\right] e^{-z/\Lambda}, \quad (4.124)$$

$$\frac{S_x^{(1)}}{|A|^2} = \frac{\omega\mu_0}{2}\text{sign}(x)\,|\sinh(p_2 d/2)|^2\,\text{Im}\left(\frac{1}{p_1}\right) e^{-(|x|-d/2)/\Lambda_1 - z/\Lambda}, \quad (4.125)$$

$$\frac{S_x^{(2)}}{|A|^2} = -\frac{\omega\mu_0}{4}\text{Im}\left[\frac{1}{p_2}\left(\sinh\left(\frac{x}{\Lambda_2}\right) - i\sin\left(\frac{4\pi x}{\lambda_2}\right)\right)\right] e^{-z/\Lambda}. \quad (4.126)$$

As for the TM modes, the in-plane and cross-plane components of the Poynting vector outside the film take its maxima at the film interfaces $x = \pm d/2$ and decay exponentially as the energy density travels away from them. Inside the film, on the other hand, this decay is not purely exponential, but the energy density still takes its maximum and minimum values at the interfaces and center ($x = 0$) of the film, respectively. In contrast to the TM modes, the thermal ($\overline{Q}_2$) and electromagnetic ($S_z^{(2)}$) energy densities inside the film exhibit the same spatial distribution and their ratio is given by

$$\frac{\overline{Q}_2}{S_z^{(2)}} = \frac{\text{Im}(\varepsilon_2)k_0^2}{\text{Re}(\beta)} \rightarrow \frac{\text{Im}(\varepsilon_2)}{\sqrt{\varepsilon_1}}k_0, \quad (4.127)$$

where the second relation applies for very thin films ($\beta_R \rightarrow k_0\sqrt{\varepsilon_1}$), as established by (4.112). The thermal-to-electromagnetic energy density of the TE even mode is thus nearly independent of the film thickness and is driven by the imaginary part of the film permittivity.

Figure 4.13a, b shows the frequency spectra of the in-plane wave vector β_R and propagation length Λ of the TE even mode, respectively. In contrast to the TM modes, this mode only propagates for frequencies outside the Reststrahlen band and the forbidden zone in red. While β_R increases with the film thickness, Λ decreases through values much smaller than the corresponding ones shown in Fig. 4.11b for the TM even mode. This short-range propagation for frequencies in which the SiC energy absorption is negligible ($\text{Im}(\varepsilon) \rightarrow 0$) indicates that the contribution of the TE even mode to the generation of heat inside the SiC film is relatively small with respect to that of the TM even mode.

TE Antisymmetric (Odd) Mode

This TE mode is defined by imposing the condition $B \neq 0$ in (4.108), which yields the following dispersion relation:

$$p_2 + p_1 \tanh\left(\frac{p_2 d}{2}\right) = 0. \quad (4.128)$$

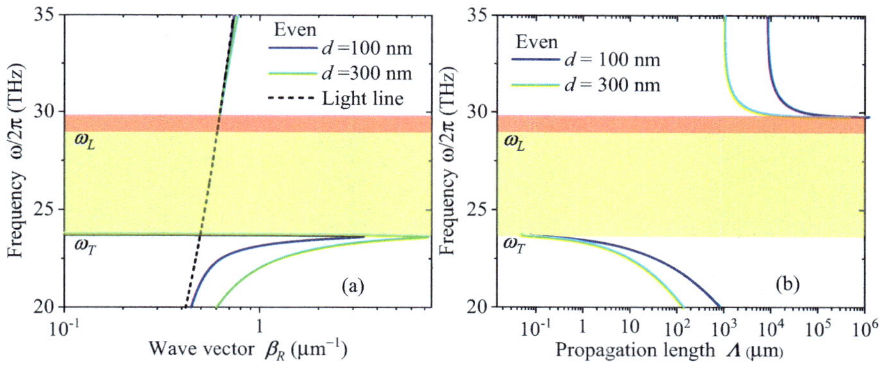

Fig. 4.13 Frequency spectrum of the in-plane (**a**) wave vector β_R and (**b**) propagation length of the even mode of the TE waves propagating along the interfaces of a SiC nanofilm suspended in vacuum. The yellow zone represents the Reststrahlen band shown in Fig. 4.3, while the red one is a forbidden band for which the dispersion relation does not have a solution

The combination of (4.128) with (4.108) and (4.109) then establishes that $A = 0$ and $A^{>} = A^{<} = B\cosh(a)$, for which the electric field inside the film is an odd $\left(E_y^{(2)}(-x) = -E_y^{(2)}(x)\right)$ function of position x. Note that for a very thin film justifying a linear approximation of the hyperbolic tangent in (4.128), we obtain $1 + p_1 d/2 = 0$, which does not have a polaritonic solution ($\mathrm{Re}(p_1) > 0$). For a not-so-thin film, on the other hand, the solution of the dispersion relation in (4.128) can be obtained by following practically the same steps as those developed for the TM odd mode in (4.84). The final results for the effective permittivity ε driving the in-plane ($\beta = k_0\sqrt{\varepsilon}$) and cross-plane ($p_n = k_0\sqrt{\varepsilon - \varepsilon_n}$) wave vectors are

$$\varepsilon = \varepsilon_1 + \left(\frac{\alpha}{h}\right)^2 \left[1 + \frac{2}{3}\frac{\alpha\,(\varepsilon_2 - \varepsilon_1)}{2\alpha + 3}h^2 + \frac{(\alpha + 1)\,(\varepsilon_2 - \varepsilon_1)^2}{\alpha\,(2\alpha + 3)^3}h^4\right], \tag{4.129}$$

where $\alpha = \sqrt[3]{(3+\sqrt{5})/2} + \sqrt[3]{(3-\sqrt{5})/2} \approx 2.1038$. As (4.129) is valid for relatively thick films ($0 \ll h < 1$), it does not generally apply for nanofilms. Furthermore, the TE odd mode may appear for very narrow ranges of frequency that usually limits its application as a generator of heat source inside films.

Given that $A = 0$ and $A^{>} = A^{<} = B\cosh(a)$, the field components in (4.103)–(4.105) for the TE odd mode can now be rewritten as follows:

$$H_z^{(1)} = B\cosh\left(p_2 d/2\right) e^{-p_1(|x| - d/2)}, \tag{4.130}$$

$$H_x^{(1)} = \frac{i\beta}{p_1} B\mathrm{sign}(x)\cosh\left(p_2 d/2\right) e^{-p_1(|x| - d/2)}, \tag{4.131}$$

$$E_y^{(1)} = \frac{i\omega\mu_0}{p_1} B\mathrm{sign}(-x)\cosh\left(p_2 d/2\right) e^{-p_1(|x| - d/2)}, \tag{4.132}$$

$$H_z^{(2)} = B\cosh\left(p_2 x\right), \tag{4.133}$$

$$H_x^{(2)} = -\frac{i\beta}{p_2} B\sinh\left(p_2 x\right), \tag{4.134}$$

$$E_y^{(2)} = \frac{i\omega\mu_0}{p_2} B\sinh\left(p_2 x\right). \tag{4.135}$$

According to (4.11), the thermal energy generated by the TE odd mode inside the thin film is determined by the electrical field in (4.135) and given by

$$\frac{\overline{Q}_2}{|B|^2} = \frac{\omega\mu_0 k_0^2}{4\,|p_2|^2}\operatorname{Im}\left(\varepsilon_2\right)\left[\cosh\left(\frac{x}{\Lambda_2}\right) - \cos\left(\frac{4\pi x}{\lambda_2}\right)\right] e^{-z/\Lambda}. \tag{4.136}$$

In addition, the Poynting vector components inside and outside the thin film can be determined by inserting (4.130) and (4.133) into (4.12) and (4.13). The final results are

$$\frac{S_z^{(1)}}{|B|^2} = \frac{\omega\mu_0}{2\,|p_1|^2}\operatorname{Re}\left(\beta\right)|\cosh(p_2 d/2)|^2\, e^{-(|x|-d/2)/\Lambda_1 - z/\Lambda}, \tag{4.137}$$

$$\frac{S_z^{(2)}}{|B|^2} = \frac{\omega\mu_0}{4\,|p_2|^2}\operatorname{Re}\left(\beta\right)\left[\cosh\left(\frac{x}{\Lambda_2}\right) - \cos\left(\frac{4\pi x}{\lambda_2}\right)\right] e^{-z/\Lambda}, \tag{4.138}$$

$$\frac{S_x^{(1)}}{|B|^2} = \frac{\omega\mu_0}{2}\operatorname{sign}(x)\,|\cosh(p_2 d/2)|^2 \operatorname{Im}\left(\frac{1}{p_1}\right) e^{-(|x|-d/2)/\Lambda_1 - z/\Lambda}, \tag{4.139}$$

$$\frac{S_x^{(2)}}{|B|^2} = -\frac{\omega\mu_0}{4}\operatorname{Im}\left[\frac{1}{p_2}\left(\sinh\left(\frac{x}{\Lambda_2}\right) + i\sin\left(\frac{4\pi x}{\lambda_2}\right)\right)\right] e^{-z/\Lambda}. \tag{4.140}$$

As for the TE even mode, the in-plane and cross-plane components of the Poynting vector for the TE odd mode take its maxima at the film interfaces $x = \pm d/2$. The energy density inside the film follows the profile of the hyperbolic cosine function modulated by the oscillations of the cosine function, as is the case of the other TE and TM modes. Furthermore, according to (4.136) and (4.138), the ratio between the thermal and electromagnetic energy densities of the TE odd mode is independent of position and given by

$$\frac{\overline{Q}_2}{S_z^{(2)}} = \frac{\operatorname{Im}(\varepsilon_2) k_0^2}{\operatorname{Re}(\beta)}. \tag{4.141}$$

The comparison between (4.127) and (4.141) indicates that even though the ratios of thermal-to-electromagnetic energy density for the even and odd modes are driven by the same propagation parameters, their absolute values are different, due to the difference of their wave vector β_R.

4.4 Cylindrical Structures

Here we consider the propagation of TM and TE waves along the surface of a circular cylinder, as shown in Fig. 4.14. The cylinder has a radius R and relative permittivity ε_1 and is surrounded by a medium of relative permittivity ε_2. Assuming that the waves are uniformly excited at the surface $z = 0$, the electric and magnetic fields become independent of the polar angle θ and depend only on the radial r and axial z coordinates. Under this angular symmetry, (3.117) reduces to the modified Bessel equation

$$\left[r^2\frac{\partial^2}{\partial r^2} + r\frac{\partial}{\partial r} - (pr)^2\right]\begin{pmatrix} E_z \\ H_z \end{pmatrix} = 0, \tag{4.142}$$

where the radial wave vector is $p^2 = \beta^2 - \varepsilon k_0^2$.

4.4.1 TM Polarization

According to (3.124)–(3.127) and (4.142), the nonzero field components (E_r, E_z, and H_θ) of the TM waves ($H_z = 0$) are given by

$$E_z = AI_0(pr) + BK_0(pr), \tag{4.143}$$

$$E_r = -\frac{i\beta}{p^2}\frac{\partial E_z}{\partial r}, \tag{4.144}$$

$$H_\theta = -\frac{i\omega\varepsilon_0\varepsilon}{p^2}\frac{\partial E_z}{\partial r}, \tag{4.145}$$

where $I_0(U)$ and $K_0(U)$ are the modified Bessel functions of order zero that grow and decay exponentially as their argument U increases, respectively. By inserting the two electric field components in (4.143) and (4.144) into (3.128), one obtains the following spectral heat source:

$$\overline{Q} = \frac{\omega\varepsilon_0}{2}\mathrm{Im}\,(\varepsilon)\left(|E_z|^2 + |E_r|^2\right)e^{-z/\Lambda}. \tag{4.146}$$

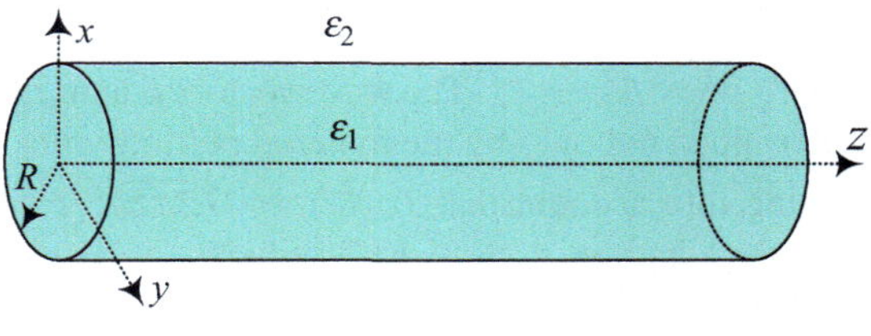

Fig. 4.14 Propagation of polaritons along the interfaces $r = R$ of a cylinder of radius R and permittivity ϵ_1

In addition, taking into account the product $\mathbf{E_0} \times \mathbf{H_0^*} = \left(E_z\hat{z} + E_r\hat{r}\right) \times H_\theta^*\hat{\theta} = \left(-E_z\hat{r} + E_r\hat{z}\right) H_\theta^*$, the axial and radial components of the Poynting vector in (3.129) take the form

$$S_z = \frac{1}{2}\text{Re}\left(E_r H_\theta^*\right) e^{-z/\Lambda} = \frac{\omega\varepsilon_0}{2\left|p\right|^4}\text{Re}\left(\beta\varepsilon^*\right)\left|\frac{\partial E_z}{\partial r}\right|^2 e^{-z/\Lambda}, \quad (4.147)$$

$$S_r = -\frac{1}{2}\text{Re}\left(E_z H_\theta^*\right) e^{-z/\Lambda} = -\frac{\omega\varepsilon_0}{2}\text{Im}\left(\frac{\varepsilon}{p^2}E_z^*\frac{\partial E_z}{\partial r}\right) e^{-z/\Lambda}. \quad (4.148)$$

The field components along with the thermal and electromagnetic energy densities are thus totally determined by parameters A and B, which can be found by imposing the continuity of the tangential field components. For the cylinder interface $r = R$ in Fig. 4.14, (3.79) and (3.80) establish that these boundary conditions read

$$E_z^{(1)} = E_z^{(2)} \qquad H_\theta^{(1)} = H_\theta^{(2)}. \quad (4.149)$$

Taking into account that $I_0(\infty) = \infty$ and $K_0(0) = \infty$, the physical (finite) solutions for the tangential fields in (4.143) and (4.145), for an arbitrary position inside and outside the cylinder shown in Fig. 4.14, can be written as follows:

$$E_z^{(1)} = AI_0(p_1 r) \quad H_\theta^{(1)} = -\frac{i\omega\varepsilon_0\varepsilon_1}{p_1}AI_1(p_1 r) \quad (4.150)$$

$$E_z^{(2)} = BK_0(p_2 r) \quad H_\theta^{(2)} = \frac{i\omega\varepsilon_0\varepsilon_2}{p_2}BK_1(p_2 r), \quad (4.151)$$

where we have used the relations $I_0'(U) = I_1(U)$ and $K_0'(U) = -K_1(U)$. After inserting (4.150) and (4.151) into (4.149) for $r = R$, one obtains

$$AI_0(p_1 R) = BK_0(p_2 R), \quad (4.152)$$

$$-\frac{\varepsilon_1}{p_1}AI_1(p_1 R) = \frac{\varepsilon_2}{p_2}AK_1(p_2 R). \quad (4.153)$$

Given that the fields are non-null ($A \neq 0 \neq B$), the division of (4.152) by (4.153) yields the following dispersion relation:

$$\frac{p_1}{\varepsilon_1}\frac{I_0(p_1 R)}{I_1(p_1 R)} + \frac{p_2}{\varepsilon_2}\frac{K_0(p_2 R)}{K_1(p_2 R)} = 0, \quad (4.154)$$

where $p_n^2 = \beta^2 - \varepsilon_n k_0^2$. The wave vectors β and p_n of the TM waves are thus strongly determined not only by the material permittivities but also by the cylinder radius R. Taking into account that $I_m(U) \approx \sqrt{2\pi U}^{-1} e^U$ and $K_m(U) \approx \sqrt{\pi/2U}e^{-U}$, for $|U| \ll 1$ and $m \geq 0$, (4.154) reduces to the usual dispersion relation of a single interface $p_1/\varepsilon_1 + p_2/\varepsilon_2 = 0$, for a thick enough cylinder ($|p_n| R \gg 1$). On the other hand, the asymptotic properties $I_0(U) \approx 1$, $I_1(U) \approx U/2$, $K_0(U) \approx -\gamma - \ln(U/2)$, and $K_1(U) \approx 1/U$, for $|U| \ll 1$ and $\gamma = 0.5772$ being the Euler–Mascheroni constant, allow simplifying (4.154) for a thin enough cylinder ($|p_n| R \ll 1$), as follows:

$$(p_2 R)^2 \left[\gamma + \ln (p_2 R/2)\right] = 2\frac{\varepsilon_2}{\varepsilon_1}. \tag{4.155}$$

The dispersion relation of the TM waves propagating along a nanowire or nanotube is thus independent of the inner radial wave vector, while its outer counterpart depends on the permittivity ratio $\varepsilon_2/\varepsilon_1$. Recalling that the solution of the equation $Ue^U = \Omega$ is given by the Lambert W function $U = W_m(\Omega)$, for $m = 0, \pm 1, \pm 2, \ldots$, the solution of (4.155) for $p_2 R$ is given by

$$(p_2 R)^2 = \frac{4\varepsilon_2/\varepsilon_1}{W_m\left[(\varepsilon_2/\varepsilon_1)\, e^{2\gamma}\right]}. \tag{4.156}$$

The solution for the axial wave vector $\beta = \sqrt{p_2^2 + \varepsilon_2 k_0^2}$ therefore takes the form

$$\beta^2 = \varepsilon_2 k_0^2 + \left(\frac{2}{R}\right)^2 \frac{\varepsilon_2/\varepsilon_1}{W_m\left[(\varepsilon_2/\varepsilon_1)\, e^{2\gamma}\right]}. \tag{4.157}$$

Considering that the Lambert W function is prebuilt in various computational software, (4.156) and (4.157) are convenient solutions for the dispersion relation of the TM waves propagating along a cylindrical nanostructure. This propagation may appear via multiple modes ($m = 0, \pm 1, \pm 2, \ldots$) that are surface polaritons in the presence of surface confinement ($\mathrm{Re}(p_2) > 0$).

The field components can now be expressed in terms of the parameter C defining $A = CK_0(p_2 R)$ and $B = CI_0(p_1 R)$ in (4.152), as follows:

$$E_z^{(1)} = CK_0(p_2 R) I_0(p_1 r), \tag{4.158}$$

$$E_r^{(1)} = -\frac{i\beta}{p_1} CK_0(p_2 R) I_1(p_1 r), \tag{4.159}$$

$$H_\theta^{(1)} = -\frac{i\omega\varepsilon_0\varepsilon_1}{p_1} CK_0(p_2 R) I_1(p_1 r), \tag{4.160}$$

$$E_z^{(2)} = CI_0(p_1 R) K_0(p_2 r), \tag{4.161}$$

$$E_r^{(2)} = \frac{i\beta}{p_2} CI_0(p_1 R) K_1(p_2 r), \tag{4.162}$$

$$H_\theta^{(2)} = \frac{i\omega\varepsilon_0\varepsilon_2}{p_2} CI_0(p_1 R) K_1(p_2 r). \tag{4.163}$$

Equations (4.158)–(4.161) thus establish that at the center of the cylinder ($r = 0$), the field components $E_r^{(1)}$ and $H_\theta^{(1)}$ vanish ($I_1(0) = 0$), while $E_z^{(1)}$ takes its minimum value ($I_0(0) = 1$). Furthermore, the combination of the electrical field components in (4.158)–(4.163) with (4.146)–(4.148) determines the following thermal and electromagnetic energy densities inside and outside the cylinder:

$$\frac{\overline{Q}_1}{|C|^2} = \frac{\omega\varepsilon_0}{2}\text{Im}\,(\varepsilon_1)\,|K_0(p_2R)|^2\left(|I_0(p_1r)|^2 + |I_1(p_1r)|^2\right)e^{-z/\Lambda}, \quad (4.164)$$

$$\frac{\overline{Q}_2}{|C|^2} = \frac{\omega\varepsilon_0}{2}\text{Im}\,(\varepsilon_2)\,|I_0(p_1R)|^2\left(|K_0(p_2r)|^2 + |K_1(p_2r)|^2\right)e^{-z/\Lambda}, \quad (4.165)$$

$$\frac{S_z^{(1)}}{|C|^2} = \frac{\omega\varepsilon_0}{2\,|p_1|^2}\text{Re}\left(\beta\varepsilon_1^*\right)|K_0(p_2R)|^2\,|I_1(p_1r)|^2\,e^{-z/\Lambda}, \quad (4.166)$$

$$\frac{S_z^{(2)}}{|C|^2} = \frac{\omega\varepsilon_0}{2\,|p_2|^2}\text{Re}\left(\beta\varepsilon_2^*\right)|I_0(p_1R)|^2\,|K_1(p_2r)|^2\,e^{-z/\Lambda}, \quad (4.167)$$

$$\frac{S_r^{(1)}}{|C|^2} = -\frac{\omega\varepsilon_0}{2}\,|K_0(p_2R)|^2\,\text{Im}\left(\frac{\varepsilon_1}{p_1}I_0^*(p_1r)I_1(p_1r)\right)e^{-z/\Lambda}, \quad (4.168)$$

$$\frac{S_r^{(2)}}{|C|^2} = \frac{\omega\varepsilon_0}{2}\,|I_0(p_1R)|^2\,\text{Im}\left(\frac{\varepsilon_2}{p_2}K_0^*(p_2r)K_1(p_2r)\right)e^{-z/\Lambda}. \quad (4.169)$$

The electromagnetic energy density ($S_z^{(1)}$ and $S_r^{(1)}$) thus vanishes at the cylinder center ($r = 0$), while the thermal one remains finite for any material and frequency with $\text{Im}(\varepsilon_n) > 0$.

4.4.2 TE Polarization

The nonzero field components (H_r, H_z, and E_θ) of the TE waves ($E_z = 0$) are given by (3.124)–(3.127) and (4.142), which yields

$$H_z = AI_0(pr) + BK_0(pr), \quad (4.170)$$

$$H_r = -\frac{i\beta}{p^2}\frac{\partial H_z}{\partial r}, \quad (4.171)$$

$$E_\theta = \frac{i\omega\mu_0}{p^2}\frac{\partial H_z}{\partial r}. \quad (4.172)$$

The electric field component in (4.172) together with (3.128) determines the following spectral heat source:

$$\overline{Q} = \frac{\omega\varepsilon_0}{2}\text{Im}\,(\varepsilon)\,|E_\theta|^2\,e^{-z/\Lambda}. \quad (4.173)$$

In contrast to the thermal energy generated by the TM waves in (4.146), the one in (4.173) for the TE waves is driven by a single component of the electric field. Furthermore, considering the product $\mathbf{E_0} \times \mathbf{H_0^*} = E_\theta \hat{\theta} \times \left(H_z^* \hat{z} + H_r^* \hat{r}\right) = E_\theta \left(H_z^* \hat{r} - H_r^* \hat{z}\right)$, the axial and radial components of the Poynting vector in (3.129) read

$$S_z = -\frac{1}{2}\mathrm{Re}\left(E_\theta H_r^*\right) e^{-z/\Lambda} = \frac{\omega\mu_0}{2\left|p\right|^4}\mathrm{Re}\left(\beta\right)\left|\frac{\partial H_z}{\partial r}\right|^2 e^{-z/\Lambda}, \tag{4.174}$$

$$S_r = \frac{1}{2}\mathrm{Re}\left(E_\theta H_z^*\right) e^{-z/\Lambda} = -\frac{\omega\mu_0}{2}\mathrm{Im}\left(\frac{H_z^*}{p^2}\frac{\partial H_z}{\partial r}\right) e^{-z/\Lambda}. \tag{4.175}$$

As for the TM waves, parameters A and B driving the fields are determined by the continuity of their tangential components at the cylinder interface $r = R$ shown in Fig. 4.14. According to (3.79) and (3.80), these boundary conditions are

$$H_z^{(1)} = H_z^{(2)} \qquad E_\theta^{(1)} = E_\theta^{(2)}. \tag{4.176}$$

Considering that $I_0(\infty) = \infty$ and $K_0(0) = \infty$, the physical solutions for these tangential components predicted by (4.170) and (4.172) for an arbitrary position r inside and outside the cylinder are

$$H_z^{(1)} = A I_0(p_1 r) \quad E_\theta^{(1)} = \frac{i\omega\mu_0}{p_1} A I_1(p_1 r). \tag{4.177}$$

$$H_z^{(2)} = B K_0(p_2 r) \quad E_\theta^{(2)} = -\frac{i\omega\mu_0}{p_2} A K_1(p_2 r). \tag{4.178}$$

The insertion of (4.177) and (4.151) into (4.176) for $r = R$ yields

$$A I_0(p_1 R) = B K_0(p_2 R), \tag{4.179}$$

$$\frac{A}{p_1} I_1(p_1 R) = -\frac{\varepsilon_2}{p_2} A K_1(p_2 R). \tag{4.180}$$

For non-null fields ($A \neq 0 \neq B$), the combination of (4.179) and (4.180) yields

$$p_1 \frac{I_0(p_1 R)}{I_1(p_1 R)} + p_2 \frac{K_0(p_2 R)}{K_1(p_2 R)} = 0. \tag{4.181}$$

Thise dispersion relation of the TE modes is similar to that of the TM modes in (4.154), which reduces to (4.181), for $\varepsilon_1 = \varepsilon_2$. Therefore, according to (4.157), the approximate analytical solution of (4.181), for a very thin cylinder ($\left|p_1\right| R \ll 1$), is given by

$$\beta^2 = \varepsilon_2 k_0^2 + \frac{(2/R)^2}{W_m\left(e^{2\gamma}\right)}. \tag{4.182}$$

Considering that $W_0\left(e^{2\gamma}\right) \approx 1.0787$ and $W_m\left(e^{2\gamma}\right)$ is a complex number for $m \neq 0$, the solutions supporting the propagation of polaritons are those for which $\Lambda = [2\mathrm{Im}(\beta)]^{-1} > 0$.

The field components for the TE waves can now be written in terms of the parameter C defining $A = CK_0(p_2R)$ and $B = CI_0(p_1R)$ in (4.179), as follows:

$$H_z^{(1)} = CK_0(p_2R)I_0(p_1r), \tag{4.183}$$

$$H_r^{(1)} = -\frac{i\beta}{p_1}CK_0(p_2R)I_1(p_1r), \tag{4.184}$$

$$E_\theta^{(1)} = \frac{i\omega\mu_0}{p_1}CK_0(p_2R)I_1(p_1r), \tag{4.185}$$

$$H_z^{(2)} = CI_0(p_1R)K_0(p_2r), \tag{4.186}$$

$$H_r^{(2)} = \frac{i\beta}{p_2}CI_0(p_1R)K_1(p_2r), \tag{4.187}$$

$$E_\theta^{(2)} = -\frac{i\omega\mu_0}{p_2}CI_0(p_1R)K_1(p_2r). \tag{4.188}$$

The field components $H_r^{(1)}$ and $E_\theta^{(1)}$ thus disappear at the center of the cylinder ($r = 0$), while $H_z^{(1)}$ takes its minimum value. The combination of the electrical field components in (4.183)–(4.188) with (4.173)–(4.175) determines the following thermal and electromagnetic energy densities inside and outside the cylinder

$$\frac{\overline{Q}_1}{|C|^2} = \frac{\omega\mu_0 k_0^2}{2|p_1|^2}\mathrm{Im}(\varepsilon_1)|K_0(p_2R)|^2|I_1(p_1r)|^2 e^{-z/\Lambda}, \tag{4.189}$$

$$\frac{\overline{Q}_2}{|C|^2} = \frac{\omega\mu_0 k_0^2}{2|p_2|^2}\mathrm{Im}(\varepsilon_2)|I_0(p_1R)|^2|K_1(p_2r)|^2 e^{-z/\Lambda}, \tag{4.190}$$

$$\frac{S_z^{(1)}}{|C|^2} = \frac{\omega\mu_0}{2|p_1|^2}\mathrm{Re}(\beta)|K_0(p_2R)|^2|I_1(p_1r)|^2 e^{-z/\Lambda}, \tag{4.191}$$

$$\frac{S_z^{(2)}}{|C|^2} = \frac{\omega\mu_0}{2|p_2|^2}\mathrm{Re}(\beta)|I_0(p_1R)|^2|K_1(p_2r)|^2 e^{-z/\Lambda}, \tag{4.192}$$

$$\frac{S_r^{(1)}}{|C|^2} = -\frac{\omega\mu_0}{2}|K_0(p_2R)|^2\,\mathrm{Im}\left(\frac{I_0^*(p_1r)}{p_1}I_1(p_1r)\right)e^{-z/\Lambda}, \tag{4.193}$$

$$\frac{S_r^{(2)}}{|C|^2} = \frac{\omega\mu_0}{2}|I_0(p_1R)|^2\,\mathrm{Im}\left(\frac{K_0^*(p_2r)}{p_2}K_1(p_2r)\right)e^{-z/\Lambda}. \tag{4.194}$$

In contrast to the thermal energy density of the TM waves, the one in (4.189) of the TE waves vanishes at the cylinder center ($r = 0$), as is the case of its electromagnetic counterparts $S_z^{(1)}$ and $S_r^{(1)}$. In fact, inside and outside the cylinder, the ratio $\overline{Q}_n/S_z^{(n)}$ is independent of position and is given by

$$\frac{\overline{Q}_n}{S_z^{(n)}} = \frac{\mathrm{Im}\,(\varepsilon_n)\,k_0^2}{\mathrm{Re}\,(\beta)} \to \frac{\mathrm{Im}\,(\varepsilon_n)\,Rk_0^2}{2\mathrm{Re}\left[\sqrt{W_m\left(e^{2\gamma}\right)}^{-1}\right]}, \tag{4.195}$$

where the second relation was obtained by taking the limit of (4.182) for $R \to 0$. In this asymptotic case, the ratio $\overline{Q}_n/S_z^{(n)}$ thus reduces linearly with the cylinder radius R and increases with the square of frequency $\omega = ck_0$.

The dispersion relation and propagation parameters of the TM and TE waves propagating along a multilayered cylindrical structure can be derived by imposing the continuity of the tangential field components at each interface and following a similar procedure than the one developed here for a single cylinder. The radii and material permittivities of those structures provide additional degrees of freedom to tune the propagation and frequency spectrum of surface polaritons [6].

References

1. Ordonez-Miranda, J., Tranchant, L., Tokunaga, T., Kim, B., Palpant, B., Chalopin, Y., Antoni, T. Volz, S.: Anomalous thermal conductivity by surface phonon-polaritons of polar nanothin films due to their asymmetric surrounding media. J. Appl. Phys. **113**, 084311 (2013)
2. Wu, Y., Ordonez-Miranda, J., Gluchko, S., Anufriev, R., De Sousa Meneses, D., Del Campo, L., Volz, S., Nomura, M.: Enhanced thermal conduction by surface phonon-polaritons. Sci. Adv. **6**, eabb4461 (2020)
3. Tranchant, L., Hamamura, S., Ordonez-Miranda, J., Yabuki, T., Vega-Flick, A., Cervantes-Alvarez, F., Alvarado-Gil, J. J., Volz, S., Miyazaki, K.: Two-dimensional phonon-polariton heat transport. Nano Lett. **19**, 6924 (2019)
4. Yang, F., Sambles, J. R., Bradberry, G. W.: Long-range surface modes supported by thin films. Phys. Rev. B **44**, 5855–5872 (1991)
5. Gluchko, S., Bruno, P., Volz, S., Braive, R., Antoni, T.: Thermal excitation of broadband and long-range surface waves on SiO2 submicron films. Appl. Phys. Lett. **110**, 263108 (2017)
6. Yeh, C., Shimabukuro, F. I.: The essence of dielectric waveguides. Springer, Berlin (2008)

Chapter 5
Polariton Heat Transport in Polar Dielectric Materials

5.1 Polariton Thermal Conductivity of Flat Structures

Let us consider a polar nanofilm supporting the simultaneous propagation of polaritons and phonons due to the temperature difference $T_h - T_c > 0$ set by two thermal baths, as shown in Fig. 5.1. The resulting steady-state heat transport along the z axis is thus driven by the combined dynamics of these two heat carriers. Assuming that the phonon heat conduction can be described by an effective thermal conductivity k_{ph}, the principle of energy conservation along with Fourier's law establishes that the temperature T inside the nanofilm is given by

$$-k_{ph}\frac{dT}{dz} + q(z) = q_t, \tag{5.1}$$

where q is the polariton heat flux and q_t is the total heat flux, a constant independent of the position z. The phonon thermal conductivity k_{ph} includes the contributions of both the acoustic phonon modes, which usually dominate the heat transport, and the optical ones that couple with photons to generate the SPhPs. Since (5.1) is a first-order differential equation, its general solution for $T(z)$ involves an integration constant, which together with q_t is determined by the boundary conditions $T(0) = T_h$ and $T(l) = T_c$, provided q is known. Considering that the steady-state heat transport appears along the $+z$ axis of the nanofilm shown in Fig. 4.1, the heat flux q is determined by its definition: the energy per unit cross-section area (ad) and per unit time that the surface polaritons carry out along the nanofilm. In terms of the polariton distribution function f, this definition of q yields [1]

$$q = \frac{1}{abd}\sum_{p}\sum_{k_y}\sum_{k_z}\hbar\omega f\mathbf{V}\cdot\hat{z}, \tag{5.2}$$

S. Volz, J. Ordonez-Miranda, *Heat Transport Driven by Surface Electromagnetic Waves*, Mechanical Engineering Series, https://doi.org/10.1007/978-3-031-66024-5_5

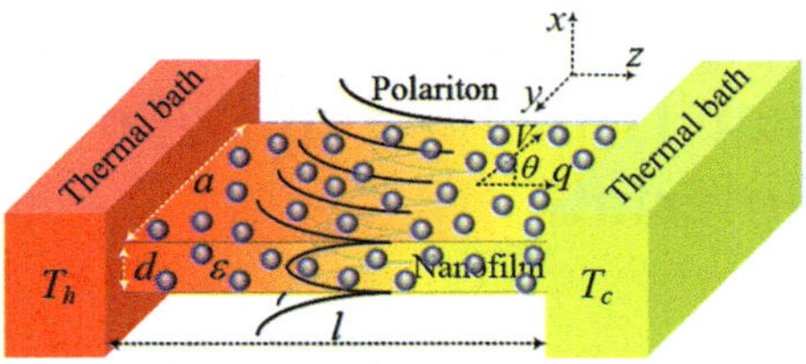

Fig. 5.1 Scheme of a polar nanofilm supporting the propagation of SPhPs due to the temperature difference $T_h - T_c > 0$ imposed by two thermal baths. The point's curvy lines represent the phonons and polaritons, respectively

where p stands for the summation over all polarizations supporting the propagation of polaritons with a group velocity $\mathbf{V}$ and wave vectors k_y and k_z in the yz plane. As in the case of electrons and phonons, the periodicity ($\psi(y + a, z) = \psi(y, z)$ and $\psi(y, z + l) = \psi(y, z)$) of the exponential wave function $\psi(y, z) \sim e^{i(k_y y + k_z z)}$ predicted by the Schrodinger equation for a single polariton establishes that $e^{i(k_y a)} = 1 = e^{i(k_z l)}$. Since $e^{2\pi n i} = 1$, for any integer n, the solutions of these relations are

$$k_y = 2\pi n/a, \quad k_z = 2\pi m/l, \quad \text{for} \quad n, m = 0, \pm 1, \pm 2, \ldots \tag{5.3}$$

The in-plane wave vectors k_y and k_z thus take only discrete values. The differences $\Delta k_y = k_y(n + 1) - k_y(n) = 2\pi/a \ll k_y(n)$ and $\Delta k_z = k_z(n + 1) - k_z(n) = 2\pi/l \ll k_z(n)$ are thus much smaller than their respective wave vectors, for the vast majority of states (large n and m integers). According to the definition of an integral, these relatively small values of Δk_y and Δk_z allow to replace the wave vectors' summations in (5.2) by integrals, as follows:

$$q = \frac{1}{(2\pi)^2 d} \sum_p \int_{-\infty}^{\infty} dk_y \int_{-\infty}^{\infty} dk_z \hbar\omega f \mathbf{V} \cdot \hat{z}. \tag{5.4}$$

Taking into account that polariton velocity makes an angle θ with the $+z$ axis (direction of q), the product $\mathbf{V} \cdot \hat{z} = V \cos(\theta)$. The presence of this angular variable indicates that the integral in (5.4) can conveniently be evaluated in polar coordinates, for which $dk_y dk_z = \beta_R d\beta_R d\theta$, with $\beta_R = \sqrt{k_y^2 + k_z^2}$ being the magnitude of the in-plane wave vectors. In terms of β_R and θ, (5.4) takes the form

$$q = \frac{1}{(2\pi)^2 d} \sum_p \int_0^{\infty} \hbar\omega\beta_R d\omega \int_0^{2\pi} f \cos(\theta) d\theta, \tag{5.5}$$

where we have used the definition of the group velocity magnitude $V = \partial\omega/\partial\beta_R$ to convert the integration from β_R to ω. The polariton heat flux is thus driven by the angular dependence of the distribution function f, which is given by the

Boltzmann transport equation (BTE). Taking into account that SPhPs are generated by the coupling of bosons (phonons and phonons), they can be treated like bosonic particles, and therefore f is determined by the bosonic BTE. Under the relaxation time approximation and the steady-state conditions imposed both the thermal baths in Fig. 5.1, this transport equation reads

$$\mathbf{V} \cdot \nabla f = \frac{f_0 - f}{\tau}, \tag{5.6}$$

where τ is the relaxation time of polaritons and f_0 is the Bose–Einstein distribution function $f_0 = \left[\exp(\hbar\omega/k_B T) - 1\right]^{-1}$. Considering that the nanofilm surfaces $z = 0$ and l are uniformly heated up by the thermal baths, the temperature variations along the x and y directions can be neglected, the temperature only varies along the z axis, and therefore the product is $\mathbf{V} \cdot \nabla f = V_z \partial f / \partial z$. Since $V_z = V \cos(\theta)$, the BTE in (5.6) can be written as follows:

$$f + \mu\Lambda \frac{\partial f}{\partial z} = f_0, \tag{5.7}$$

where $\mu = \cos(\theta)$ and $\Lambda = V\tau$ is the polariton propagation length. Considering that f_0 is independent of θ, (5.7) establishes that the angular dependence of f shows up via μ only ($f = f(\mu)$). Therefore, the angular integral in (5.5) can conveniently be split into two parts, as follows:

$$\int_0^{2\pi} f \cos(\theta) d\theta = \int_{-\pi/2}^{\pi/2} f(\cos(\theta)) \cos(\theta) d\theta + \int_{\pi/2}^{3\pi/2} f(\cos(\theta)) \cos(\theta) d\theta. \tag{5.8}$$

The first and second integrals in the right-hand side of (5.8) represent the contributions of the polaritons propagating from left to right ($-\pi/2 \leq \mu \leq \pi/2$) and right to left ($\pi/2 \leq \mu \leq 3\pi/2$), respectively. The change of variables $\phi = \theta - \pi$ ($\cos(\theta) = -\cos(\phi)$) transforms the integrant of the second integral in the right-hand side of (5.8) into $-f(-\cos(\phi)) \cos(\phi)$, which change in the interval $-\pi/2 \leq \phi \leq \pi/2$. In terms of the dummy variable μ, (5.8) takes the form

$$\int_0^{2\pi} \mu f d\theta = \int_{-\pi/2}^{\pi/2} \mu \left[f(\mu) - f(-\mu)\right] d\theta. \tag{5.9}$$

Since the integrant in the right-hand side integral of (5.9) is an even function of θ ($\cos(-\theta) = \cos(\theta)$), its symmetric bounds around zero allow to rewrite the polariton heat flux in (5.5), as follows:

$$q = \frac{1}{2\pi^2 d} \sum_p \int_0^{\infty} \hbar\omega\beta_R F d\omega, \tag{5.10}$$

$$F = \int_0^{\pi/2} \mu \left[f(\mu) - f(-\mu) \right] d\theta. \tag{5.11}$$

To better understand the predictions of (5.10) and (5.11), we are going to consider the following limiting cases:

5.1.1 Diffusive Heat Transport: $l \gg \Lambda$

A diffusion approximation is valid when the nanofilm length is several times longer than the polariton in-plane propagation length ($l \gg \Lambda$). This condition is analogous to the Rosseland diffusion approximation used in thermal radiation and implies the existence and propagation of polaritons near thermal equilibrium (small temperature gradients). Therefore, the polariton distribution function does not deviate significantly from its equilibrium counterpart, and its position variations can be approximated by $\partial f / \partial z \approx \partial f_0 / \partial z$. Under this first-order approximation, the BTE in (5.7) is said to be linearized and yields the following distribution function:

$$f = f_0 - \mu \Lambda \frac{\partial f_0}{\partial z}. \tag{5.12}$$

The difference $f(\mu) - f(-\mu) = -2\mu\Lambda\partial f_0/\partial z$ thus depends on the equilibrium distribution function f_0 through its derivative only. Inserting (5.12) into (5.11), one obtains

$$F = -2\Lambda \frac{\partial f_0}{\partial z} \int_0^{\pi/2} \cos^2(\theta) d\theta = -\frac{\pi}{2} \Lambda \frac{\partial f_0}{\partial z}. \tag{5.13}$$

Taking into account that f_0 depends on position z via the temperature $T = T(z)$ only, the chain rule $\partial f_0/\partial z = (\partial f_0/\partial T)(\partial T/\partial z)$ along with Fourier's law $q = -k\partial T/\partial z$ and (5.10) predicts the following polariton thermal conductivity k:

$$k = \frac{1}{4\pi d} \sum_p \int_0^{\infty} \hbar\omega \beta_R \Lambda \frac{\partial f_0}{\partial T} d\omega. \tag{5.14}$$

Equation (5.10) was first derived by Gang Chen group [2], for a single mode, and is analogous to the well-phonon formula for the phonon thermal conductivity derived for a three-dimensional density of states [1]. However, in contrast to its phononic counterpart, k is independent of the polariton group velocity due to the two-dimensional density of states $\left[\beta_R/(2\pi)^2 V\right]$ of polaritons [2, 3]. The polariton thermal conductivity of a nanofilm supporting the diffusive propagation of polaritons thus depends on the product ($\beta_R \Lambda$) between the in-plane wave vector and propagation length, which are determined by the polariton dispersion relation for the p =TM and TE polarizations, as discussed in Chap. 4. The particle description of polaritons through the BTE thus involves parameters of their wave description driven by Maxwell's equations.

5.1.2 Ballistic Heat Transport: $l \ll \Lambda$

The ballistic limit appears when the nanofilm length is much smaller than the in-plane propagation length ($l \ll \Lambda$). Under this condition of long propagation length, $(f_0 - f)/\Lambda \to 0$, and therefore (5.7) reduces to

$$\mu \frac{\partial f}{\partial z} = 0. \tag{5.15}$$

Considering the boundary conditions $T(z = 0) = T_h$ and $T(z = l) = T_c$ and the emission of polaritons from both nanofilm surfaces $z = 0$ and $z = l$ in thermal equilibrium with the thermal baths, the solution of (5.15) is

$$f = \begin{cases} f_0(T_h) & \text{for} \quad 0 < \mu < 1 \ (0 < \theta < \pi/2), \\ f_0(T_c) & \text{for} \quad -1 < \mu < 0 \ (-\pi/2 < \theta < 0). \end{cases} \tag{5.16}$$

The combination of (5.15) with (5.11) yields

$$F = [f_0(T_h) - f_0(T_c)] \int_0^{\pi/2} \cos(\theta) d\theta = f_0(T_h) - f_0(T_c). \tag{5.17}$$

Considering that the temperature difference $\Delta T = T_h - T_c$ is much smaller than average temperature $T = (Th + Tc)/2$, the evaluation of f_0 at $T_h = T + \Delta T/2$ and $T_c = T - \Delta T/2$ can be done via the thermal series expansion, whose first-order approximation is

$$f_0(T \pm \Delta T/2) \approx f_0(T) \pm \frac{\Delta T}{2} \frac{\partial f_0(T)}{\partial T}. \tag{5.18}$$

Equations (5.17) and (5.18) thus establish that $F = \Delta T \partial f_0/\partial T$, which together with (5.10) predicts that the polariton thermal conductivity $k = ql/\Delta T$ of a nanofilm supporting the ballistic propagation of polaritons is

$$k = \frac{l}{2\pi^2 d} \sum_p \int_0^\infty \hbar \omega \beta_R \frac{\partial f_0}{\partial T} d\omega. \tag{5.19}$$

In the ballistic regime, the polariton thermal conductivity is thus driven by the in-plane wave vector β_R and no longer depends on the propagation length Λ, as in the case of the diffusive limit in (5.14). The comparison of (5.14) and (5.19) indicates that the ballistic and diffusive polaritons generate the same thermal conductivity, when their propagation length $\Lambda = 2l/\pi$ becomes independent of frequency and is totally determined by the film length. In terms of the thermal conductance $G = qad/\Delta T = kad/l$, which is a convenient metric for the ballistic heat transport, (5.19) reads

$$\frac{G}{a} = \frac{1}{2\pi^2} \sum_p \int_0^\infty \hbar\omega\beta_R \frac{\partial f_0}{\partial T} d\omega. \tag{5.20}$$

Considering that, for a thin enough nanofilm, the even mode of the TM waves propagates long distances ($\Lambda \gg l$) with a wave vector $\beta_R = \sqrt{\varepsilon_1}\omega/c$ independent of the nanofilm properties, as established by (4.68) and (4.69), the contribution G_0 of this long-range mode to the ballistic polariton thermal conductance is

$$\frac{G_0}{a\sqrt{\varepsilon_1}} = \frac{\hbar}{2c\pi^2} \int_0^\infty \omega^2 \frac{\partial f_0}{\partial T} d\omega = \frac{2k_B^3 T^2}{ch^2} \int_0^\infty \frac{x^3 e^x}{(e^x - 1)^2} dx = 12z(3)\frac{k_B^3 T^2}{ch^2}, \tag{5.21}$$

where $x = \hbar\omega/k_B T$, $z(3) = \sum_{n=1}^\infty n^{-3} \approx 1.2021$, and we assumed that the relative permittivity ε_1 of the medium surrounding the nanofilm is independent of frequency. Therefore, when the ballistic propagation of the symmetric TM mode occurs for all frequencies, its thermal conductance per unit width a and per $\sqrt{\varepsilon_1}$ is independent of the nanofilm properties, depends on universal constants, and increases with the square of temperature. At room temperature ($T = 300\,\mathrm{K}$), $G_0/a\sqrt{\varepsilon_1} = 25.94\,\mathrm{nW\,mm^{-1}\,K^{-1}}$, which increases to $G_0/a\sqrt{\varepsilon_1} = 72.05\,\mathrm{nW\,mm^{-1}\,K^{-1}}$, for $T = 500\,\mathrm{K}$. This quantization of the polariton thermal conductance of a nanofilm is consistent with the T^3 increase of the 2D radiative heat flux discussed in Sect. 2.1.4 and analogous to its phonon and electron counterparts predicted for 2D structures at low temperatures [4–6]. Furthermore, (5.21) was also used to describe the enhancement of the measured far-field thermal radiation between silicon microplates coated with SiO_2 nanolayers supporting the excitation and propagation of SPhPs [7].

5.1.3 *Diffusive-Ballistic Heat Transport*

In this regime, the polariton propagation length can be longer or smaller than the nanofilm length, and therefore the distribution function f does not have a simple expression, as in the previous cases for pure diffusive or ballistic polaritons. However, in terms of the normalized position $\xi = z/\Lambda$, the general solution of the BTE in (5.7) can still be obtained analytically and given by

$$f(\xi, \mu) = f(\xi_0, \mu)e^{(\xi_0-\xi)/\mu} + \frac{1}{\mu}\int_{\xi_0}^{\xi} f_0(\xi')e^{(\xi'-\xi)/\mu} d\xi', \tag{5.22}$$

where ξ_0 is an arbitrary constant and $f_0(\xi') \equiv f_0(T(\xi'))$. Since the film surfaces $z = 0$ and l are in thermal equilibrium with the corresponding thermal baths, they are subject to the following boundary conditions:

$$f(0, \mu) = f_0(0) \quad \text{for} \quad 0 < \mu < 1, \tag{5.23}$$

$$f(\lambda, \mu) = f_0(\lambda) \quad \text{for} \quad -1 < \mu < 0, \tag{5.24}$$

where $\lambda = l/\Lambda$. To facilitate the evaluation of these boundary conditions, Eq. (5.11) is conveniently split it into two parts, as follows:

$$f^+(\xi, \mu) = f_0(0)e^{-\xi/\mu} + \frac{1}{\mu}\int_0^{\xi} f_0(\xi')e^{(\xi'-\xi)/\mu}d\xi', \tag{5.25}$$

$$f^-(\lambda, \mu) = f_0(\lambda)e^{(\lambda-\xi)/\mu} - \frac{1}{\mu}\int_{\xi}^{\lambda} f_0(\xi')e^{(\xi'-\xi)/\mu}d\xi', \tag{5.26}$$

where the superscripts "+" and "−" stand for $0 < \mu < 1$ and $-1 < \mu < 0$, respectively. This split of the distribution functions thus seamlessly integrates the boundary conditions in (5.23) and (5.24). By inserting (5.25) and (5.26) into (5.11), one obtains

$$\begin{aligned} F &= \int_0^{\pi/2} \mu\left[f^+(\xi, \mu) - f^-(\xi, -\mu)\right]d\theta \\ &= f_0(0)E_3(\xi) - f_0(\lambda)E_3(\lambda - \xi) - \frac{d}{d\xi}\int_0^{\lambda} f_0(\xi')E_3\left(\left|\xi' - \xi\right|\right)d\xi', \end{aligned} \tag{5.27}$$

where, by analogy with the exponential integral function, $E_n(x)$ is defined by

$$E_n(x) = \int_0^{\pi/2} \mu^{n-2}e^{-x/\mu}d\theta, \tag{5.28}$$

for $\mu = \cos(\theta)$. The function E_n varies between zero and unity for any $n = 1, 2, \ldots$. For the sake of simplicity, we rewrite (5.27) in terms of the normalized distribution function $g = (f_0(\xi) - f_0(\lambda)) \,/\, (f_0(0) - f_0(\lambda))$, as follows:

$$F = (f_0(0) - f_0(\lambda))\left[E_3(\xi) - \frac{d}{d\xi}\int_0^{\lambda} g(\xi')E_3\left(\left|\xi' - \xi\right|\right)d\xi'\right]. \tag{5.29}$$

Assuming the nanofilm temperature difference $\Delta T = T_h - T_c \ll (T_h + T_c)/2 = \bar{T}$, the distribution difference $f_0(0) - f_0(\lambda) \equiv f_0(T_h) - f_0(T_c) = \Delta T \partial f_0(\bar{T})/\partial \bar{T}$ is well approximated by the linear expansion of the Taylor series. Under this assumption, the temperature difference $\left|T(\xi) - \bar{T}\right| \ll \bar{T}$, for any position ξ inside the nanofilm, and therefore $f_0(\xi) \equiv f_0(T(\xi)) = f_0\left[\bar{T} + (T(\xi) - \bar{T})\right] = f_0(\bar{T}) + (T(\xi) - \bar{T})\partial f_0(\bar{T})/\partial \bar{T}$ can also be linearized on $T(\xi) - \bar{T}$. Furthermore, the small temperature gradients of the nanofilm are also expected to generate a nearly linear dependence of its temperature on position ($f_0(T(\xi)) \sim T(\xi) \sim \xi$). Therefore, the linear approximation of the equilibrium intensity on position can be written as $f_0(\xi) = \alpha(\beta - \xi)$, with α and β being two parameters independent of position. Under this condition, $g(\xi) = 1 - \xi/\lambda$ and (5.29) takes the form

$$F = \frac{\Delta T}{\lambda}\frac{\partial f_0}{\partial T}\left[\frac{\pi}{2} - E_4(\xi) - E_4(\lambda - \xi)\right], \tag{5.30}$$

where we have replaced $\bar{T}$ by T, for the sake of simplicity. In contrast to the diffusive and ballistic regimes, F depends not only f_0 but also on position through the argument $\xi = z/\Lambda$ of function E_4. As the propagation length Λ generally varies with frequency, the frequency dependence of F arises from both f_0 and E_4. According to (5.10), (5.11), and (5.30), the polariton heat flux is determined by the following integral on frequency:

$$q(z) = \frac{\Delta T}{2\pi^2 d}\sum_p \int_0^\infty \frac{\hbar\omega\beta_R}{\lambda}\frac{\partial f_0}{\partial T}\left[\frac{\pi}{2} - E_4(\xi) - E_4(\lambda - \xi)\right] d\omega. \tag{5.31}$$

The position dependence of q indicates that the polaritons and phonons inside the nanofilm are coupled and exchange their energy, such that their combined heat flux remains constant, as established by the conservation of energy in (5.1). Note that the polariton heat flux at two equidistant positions from the external film surfaces ($z = 0; \lambda$) takes the same value ($q(z) = q(l - z)$), such that its maximum appears at the middle of the film ($z = l/2$). This behavior is reasonable, as the trade-off between the local temperature gradient and polariton population is expected to be maximum at the center of the film. Taking into account the property $E_n'(x) = -E_{n-1}(x)$ derived from the definition of E_n in (5.28), the integration of (5.1) for the polariton heat flux in (5.31) yields the following expressions for the total (phonon+polariton) heat flux and temperature:

$$q_t = \left(k + k_{ph}\right)\frac{\Delta T}{l}, \tag{5.32}$$

$$T(z) = T_h - \Delta T\frac{z}{l} + \frac{2}{\pi}\frac{\Delta T}{k_{ph}}\int_0^\infty k_\omega \left[\frac{\psi(z/\Lambda) + (2z/l - 1)\,\psi(0)}{\lambda - 4\psi(0)/\pi}\right] d\omega, \tag{5.33}$$

where $\Delta T = T_h - T_c$, $\psi(\xi) = E_5(\xi) - E_5(\lambda - \xi)$, and the spectral polariton thermal conductivity k_ω is defined by its integrated counterpart ($k = \int k_\omega d\omega$) given by

$$k = \frac{1}{4\pi d}\sum_p \int_0^\infty \hbar\omega\beta_R\Lambda_e\frac{\partial f_0}{\partial T} d\omega, \tag{5.34}$$

$$\Lambda_e = \Lambda\left(1 - \frac{4\psi(0)}{\pi\lambda}\right). \tag{5.35}$$

The formula for the polariton thermal conductivity k has therefore the same mathematical form as the one derived for the diffusive regime, with the intrinsic propagation length Λ replaced by the effective one Λ_e in (5.35). Λ_e is determined by the ratio $\lambda = l/\Lambda$ between the film length l and Λ, and therefore it captures the film length effects. In the diffusive limit ($\lambda = l/\Lambda \gg 1$), the ratio $\psi(0)/\lambda \rightarrow 0$ and $\Lambda_e = \Lambda$, as expected. In the ballistic limit ($\lambda \ll 1$), on the other hand, $1 - 4\psi(0)/\pi\lambda \approx 2\lambda/\pi$, $\Lambda_e = 2l/\pi$, and (5.34) reduces to the ballistic polariton thermal conductivity in (5.19). Equations (5.34) and (5.35) therefore unify the separate results obtained for diffusive or ballistic polaritons and provide the polariton thermal conductivity of a nanofilm with an arbitrary length.

The effective propagation length in (5.35) is predicted by the BTE and is different than Matthiessen's rule (MR) $\Lambda_e = \Lambda l/(\Lambda + l)$ commonly used for describing the effective mean free path of phonons. The semiempirical MR sums over the intrinsic $(1/\Lambda)$ and boundary $(1/l)$ scattering rates in a phenomenological way, while the BTE seamlessly integrates the boundary scattering through the boundary conditions used to obtain its solution. For the ballistic ($l \ll \Lambda$) and ballistic-diffusive regimes ($l \sim \Lambda$), the MR overestimates the effective propagation length predicted by the BTE, as shown in Fig. 5.2. By contrast, in the diffusive regime ($l \gg \Lambda$), the deviation of the MR predictions from the BTE ones is small. Taking into account that the BTE solution was rigorously derived without resorting to the semiempirical concepts, the predictions of (5.35) are thus expected to be more accurate than those of the MR, especially in the ballistic and ballistic-diffusive regimes. These two regimes are expected to be the ones of practical interest, as the long-range polaritons propagate long distances of up to a few centimeters along a nanofilm, while its length typically takes comparable or smaller values.

As the heat transport in the nanofilm is driven by the combined dynamics of polaritons and phonons, the total heat flux is determined by the sum $(k + k_{ph})$ of thermal conductivities related to these two energy carriers, as established by (5.32). The usual reduction of k_{ph}, as the film thickness scales down, could therefore

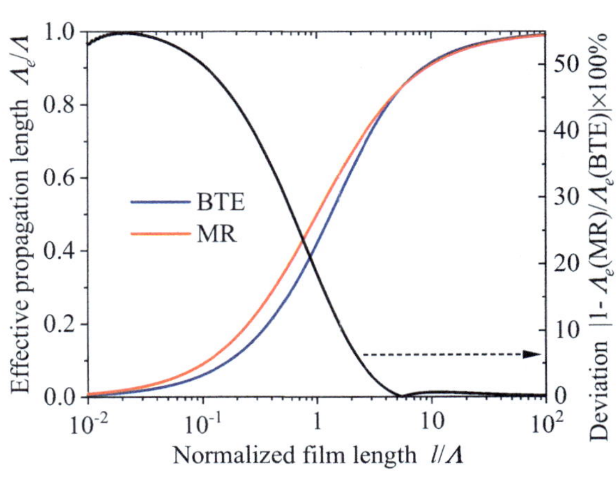

Fig. 5.2 Effective propagation length of polaritons predicted by the BTE and Matthiessen's rule. The black line stands for the deviation of the predictions of this rule

be offset by the increasing values of k, due to the predominant surface effects driving the propagation of polaritons. According to (5.33), the polariton propagation also changes the spatial distribution of temperature, whose deviation from the usual linear profile (first two terms) is driven by the ratio k/k_{ph}. Interestingly, regardless of the values of this ratio, the polariton contribution to the temperature profile disappears not only at the external surfaces $z = 0$ and l but also at the middle of the film ($z = l/2$). This behavior is related to the symmetry of the polariton heat flux ($q(z) = q(l - z)$) around $z = l/2$, as the sum of temperatures $T(z) + T(l - z) = T_h - T_c$ is an invariant of heat conduction, as predicted by (5.33). This feature of temperature is generated by the nonlocal behavior (dependence on z and $l - z$ of q in (5.31)) of the heat conduction and is analogous to the characteristic temperature profiles found in radiative heat transfer [8].

The polariton contribution to the heat transport along the nanofilm can also be characterized by its polariton thermal conductance $G = adk/l$, which according to (5.34) is given by

$$\frac{G}{a} = \frac{1}{2\pi^2} \sum_p \int_0^\infty \hbar\omega\beta_R\tau \frac{\partial f_0}{\partial T} d\omega, \tag{5.36}$$

$$\tau = \frac{\pi}{2l}\Lambda_e = \frac{\pi}{2\lambda}\left(1 - \frac{4\psi(0)}{\pi\lambda}\right). \tag{5.37}$$

The polariton propagation of heat along the nanofilm can thus be viewed as a transmission process characterized by the transmission probability τ (transmittivity) driven by $\lambda = l/\Lambda$. In the diffusive regime ($\lambda \gg 1$), $\tau \approx 0$, and the polariton thermal conductance vanishes ($G \rightarrow 0$). By contract, in the ballistic regime ($\lambda \ll 1$), $\tau \approx 1$, and (5.36) reduces to (5.20), as expected. The polariton thermal conductance is hence enhanced along a film with length smaller than the polariton propagation length and depends on the film properties through the product $\beta_R\tau = \mathrm{Re}(\beta)\tau(2l/\mathrm{Im}(\beta))$ driven by the real and imaginary parts of the polariton wave vector β. The frequency dependence of $\beta(\omega)$ is determined by the dispersion relation of polaritons propagating along the nanofilm shown in Fig. 5.1. As G increases with this product, the optimal material configuration to maximize the polariton heat transport is given by a large wave vector $\mathrm{Re}(\beta)$ and a long propagation length (small $\mathrm{Im}(\beta)$). (5.36) has the mathematical form of Landauer's formula for the phonon or electron thermal conductance of films and wires [9], and therefore we call it Landauer's formula for the polariton thermal conductance.

5.1.4 Graphical Analysis

It is interesting to note that, even though the formulas for the thermal conductivity k and thermal conductance G of polaritons were derived by treating the polaritons like particles via the BTE, they depend on the wave nature of polaritons, through their in-

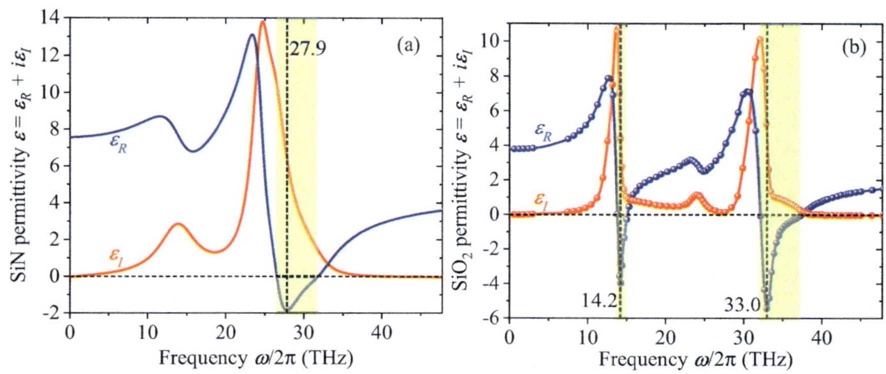

Fig. 5.3 Real (ε_R) and imaginary (ε_I) parts of the relative permittivity of (**a**) SiN [10] and (**b**) SiO_2 [11, 12], as functions of frequency. The yellow zones represent the frequency windows in which $\varepsilon_R < 0$ (Reststrahlen band)

plane wave vector β. In nonmagnetic polar materials, the real and imaginary parts of β are driven by their permittivity, as shown in Chap. 4. We now quantify the values of k and G for suspended nanofilms ($\varepsilon_1 = 1$) of SiC, SiN, and SiO_2, three common polar materials. Figure 4.3 shows the real (ε_R) and imaginary (ε_I) parts of the relative permittivity of SiC, while those of SiN and SiO_2 are shown in Fig. 5.3a, b, respectively. Note that the three materials absorb energy ($\varepsilon_I > 0$) and favor the surface confinement of polaritons ($\varepsilon_R < 0$) in narrowbands of frequency. While SiC and SiN exhibit one major absorption peak and one Reststrahlen band (yellow zones) only, SiO_2 has two of them. The combined effects of both $\varepsilon_R < 0$ and $\varepsilon_I < 0$ drive the frequency spectrum of the intrinsic propagation length $\Lambda = [2\text{Im}(\beta)]^{-1}$, which tends to take relatively small values at the absorption peaks, as shown in Fig. 5.4. Both nanofilms of SiC and Si_2 have forbidden bands (reddish zones) in which the polariton dispersion relation does not have solution, and therefore polaritons do not propagate. By contrast, the SiN nanofilm supports the polariton propagation for all frequencies. Furthermore, the increase of Λ for sufficiently low and high frequencies is generated by the weak absorption ($\varepsilon_I \to 0$), such that longer Λ values are obtained for smaller ε_I values. This low-frequency and high-frequency increase of Λ is, however, expected to not contribute significantly to the polariton heat transport due to the weak absorption of the polariton energy by the material. On the other hand, for the three considered polar materials, the real part of the in-plane wave vector is well approximated by the light line $\beta_R = k_0 = \omega/c$, as shown in Fig. 4.11, for the TM even mode in a SiC nanofilm. This approximation remains accurate for nanofilms thinner than 100 nm and frequencies $\omega/2\pi < 100$ THz.

Figure 5.5 shows the frequency spectrum (k_ω) of the polariton thermal conductivity spectrum determined by (5.34) and the propagation length shown in Fig. 5.4, for nanofilms of SiN, SiC, and SiO_2. Note that, for these three materials, k_ω is pretty much the same for the vast majority of frequencies and temperatures up to 500 K. Apart from the minor differences due to permittivity particularities and the

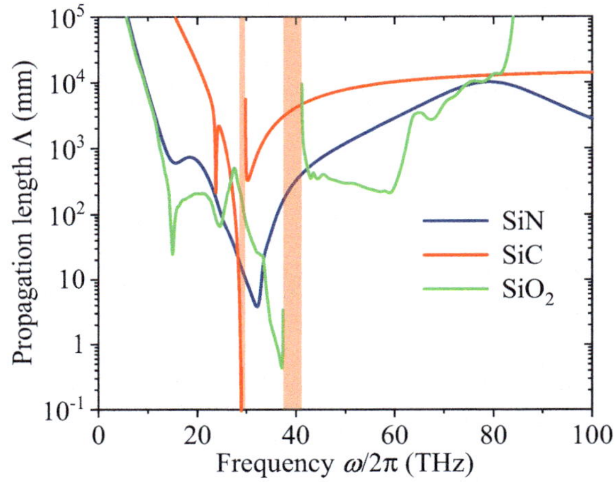

Fig. 5.4 Frequency spectrum of the propagation length of the long-range polaritons (TM even mode) propagating along a 50-nm-thick nanofilm of SiN, SiC, and SiO_2. The reddish zones represent the forbidden zones, where the polariton dispersion relation does not have a solution and hence polaritons do not propagate

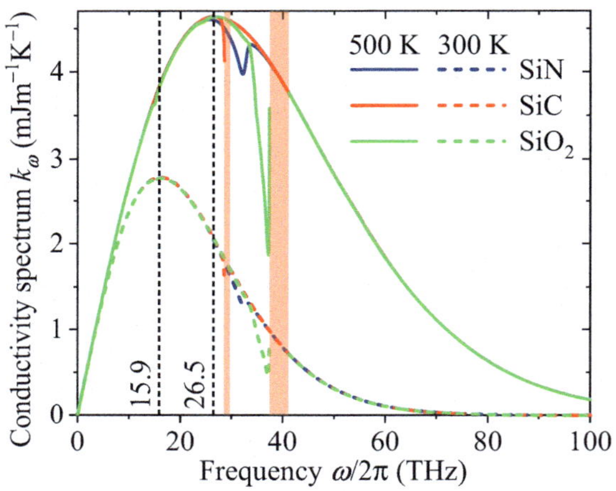

Fig. 5.5 Frequency spectrum of the polariton thermal conductivity of a 50-nm-thick nanofilm of SiN, SiC, and SiO_2 at 300 and 500 K. Calculations were done for the contribution of the TM even mode and a nanofilm length $l = 1$ mm

forbidden bands in the SiC and SiO_2 nanofilms, this behavior independent of the material is due to the quasiballistic propagation of polaritons ($\Lambda \gg l$), for the considered nanofilm length $l = 1$ mm. The frequency and amplitude of the peak of k_ω increase with temperature, as established by the Bose–Einstein distribution function f_0.

The integrated thermal conductivity k of polaritons propagating along a polar nanofilm is shown in Fig. 5.6a–c, for nanofilm lengths of $l = 100\,\mu$m, 1 mm, and 1 cm, respectively. For the shortest film ($l = 100\,\mu$m), the values of k are practically independent of the nanofilm material due to the pure ballistic propagation ($\Lambda \gg l$) of polaritons with an effective propagation length $\Lambda_e = 2l/\pi$. For the longer nanofilms, polaritons have a quasiballistic propagation ($\Lambda \approx l$), and therefore the material difference has an impact on k. The values of k increase as the temperature and the length of the nanofilm increase or its thickness decreases. In particular, note that the polariton thermal conductivity of a 1-mm-long nanofilm of SiO_2 can be

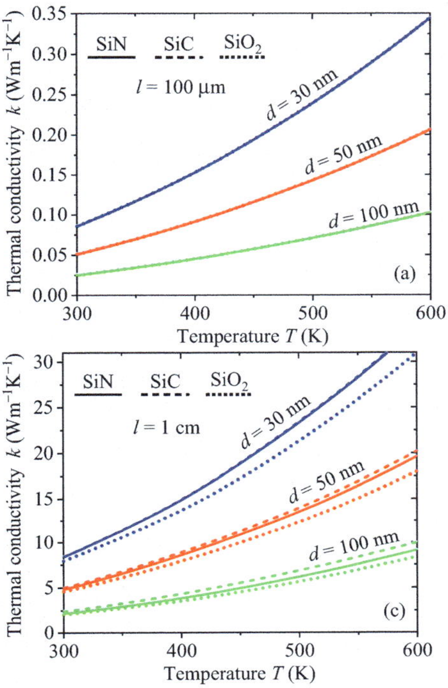

Fig. 5.6 Polariton thermal conductivity of a nanofilm of SiN, SiC, and SiO_2 as a function of temperature. Calculations were done for three nanofilm thicknesses, for each nanofilm length of (**a**) 100 μm, (**b**) 1 mm, and (**c**) 1 cm

higher than its phonon counterpart ($k > k_{ph} = 1.4\,\mathrm{Wm}^{-1}\mathrm{K}^{-1}$), for a thickness $d < 50$ nm and temperature $T > 300$ K. These facts reveal the great potential of surface phonon–polaritons to enhance the heat conduction along polar nanofilms. This enhancement is expected to become significant at high temperatures, for which the polariton thermal conductivity is important, while its phonon counterpart is usually small in polar dielectric materials.

Figure 5.7a shows the measured in-plane thermal conductivity (k_{exp}) of amorphous SiN membranes with thicknesses of 30, 50, 100, and 200 nm, for temperatures ranging from 300 to 800 K. For the 200-nm thick membrane at 300 K, $k_{exp} = 4.36$ W $\mathrm{m}^{-1}\mathrm{K}^{-1}$ and remains nearly temperature independent up to 500 K. Above 500 K, the thermal conductivity decreases according to the power law $k_{exp} \propto T^{-2}$. This high-temperature reduction is known to be caused by the internal phonon scattering processes, which indicates that heat conduction in this relatively thick membrane is driven by phonons mainly. By contrast, the thermal conductivity of the 100-nm-thick membrane is nearly independent of temperature, with the values of 1.25 and 0.99 W $\mathrm{m}^{-1}\mathrm{K}^{-1}$ at 300 and 800 K, respectively. This lack of sensitivity on temperature is explained by the increasing contribution of polaritons that

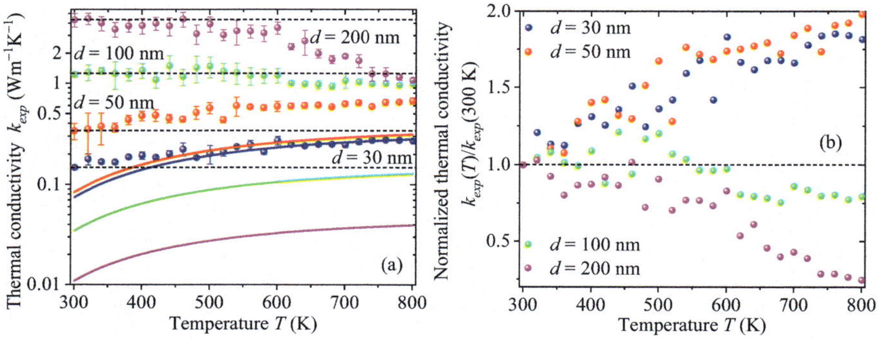

Fig. 5.7 (**a**) Absolute and (**b**) normalized values of the measured in-plane thermal conductivity of SiN membranes as a function of temperature. The solid lines represent the polariton thermal conductivity predicted by (5.34) for the four SiN membranes [10]. Enhancement of heat conduction is observed for the 30- and 50-nm-thick membranes at high temperature

compensates the reduction of the phonon thermal conductivity, as the temperature rises. For the two thinner membranes ($d \leq 50$ nm), on the other hand, the increase of k_{exp} with temperature becomes notable, so that its value rises from 0.34 Wm^{-1}K^{-1} at 300 K to 0.68 W m^{-1}K^{-1} at 800 K, for the 50-nm-thick membrane. The thermal conductivity of the 30-nm-thick membrane exhibits a similar enhancement with values of 0.15 and 0.27 Wm^{-1}K^{-1} at 300 and 800 K, respectively. The temperature evolution of the thermal conductivity of the four membranes becomes even more evident when the values are normalized by the corresponding values at room temperature, as shown in Fig. 5.7b. The two thinner SiN membranes become twice more conductive as their temperature rises from 300 to 800 K. This increase of thermal conductivity with temperature cannot be explained by mechanisms other than polaritons, which reveals their potential as an effective heat carrier for enhancing the heat dissipation in nanomaterials.

Taking into account that the SiN membranes support the propagation of both phonons and polaritons, the measured thermal conductivity $k_{exp} = k(\text{phonon}) + k(\text{SPhP})$ is determined by the phonon and polariton contributions. For the thinnest membrane ($d = 30$ nm), the values of the measured thermal conductivity agree with those predicted for its polariton counterpart ($k_{exp} \approx k(\text{SPhP})$), for temperatures higher than 500 K, where the phonon contribution is expected to be relatively small. For lower temperatures, on the other hand, the latter contribution does matter and hence $k_{exp} > k(\text{SPhP})$. Furthermore, the predicted values of $k(\text{SPhP})$ (blue line in Fig. 5.7a) for the 30-nm-thick membrane are lower than the ones for a 50-nm-thick membrane due to its relatively small length, which reduces its effective propagation length. While the thinnest membrane has a length $l = 0.5$ mm, the others have the same length $l = 1$ mm. The remarkable difference between k_{exp} and $k(\text{SPhP})$ for the three thicker membranes is explained by the dominant phonon contribution $k(\text{phonon})$, which increases with the membrane thickness.

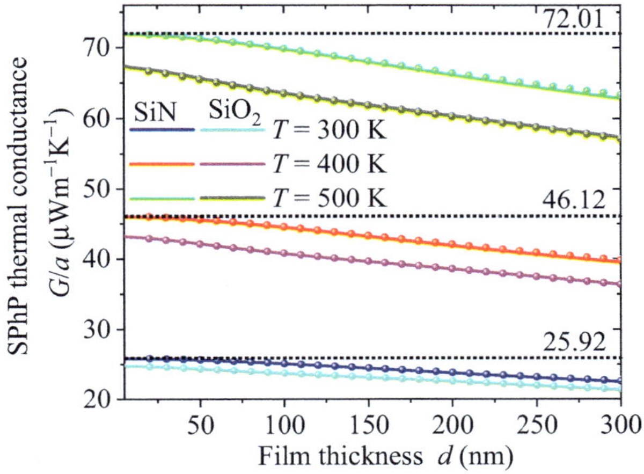

Fig. 5.8 Polariton thermal conductance per unit width of 1-mm-long nanofilms of SiN and SiO_2, as a function of their thickness. The solid lines and dots represent the predictions of the respective analytical (5.36) and numerical [3] solutions of the BTE, while the black dotted lines stand for the quantum of thermal conductance predicted by (5.21). Calculations were done for three representative temperatures

The polariton thermal conductance per unit width G/a of SiN and SiO_2 nanofilms is comparatively shown in Fig. 5.8, as a function of the film thickness. As a result of the predominance of surface effects, thinner films have a higher G/a, whose values increase with temperature. The analytical (solid lines) and numerical (dashed lines) predictions exhibit a very good agreement for both materials and three temperatures, which confirms the high accuracy of (5.36) for quantifying the SPhP thermal conductance. More importantly, in the ballistic regime ($d < 50$ nm), the upper bound of G/a for a SiN nanofilm coincides with the quantum of thermal conductance in (5.21), for the three considered temperatures. This agreement is due to the fact that SiN is able to support the polariton propagation for all relevant frequencies, without a forbidden band. By contrast, the frequency band gap of SiO_2 (Fig. 5.4) reduces its polariton thermal conductance to values slightly smaller than the quantum of thermal conductance. This quantization can thus only be achieved in nanofilms without frequency band gaps for the propagation of SPhPs.

Figure 5.9a, b shows the respective temperature T and heat flux q profiles predicted by the BTE for the simultaneous propagation of polaritons and phonons in a 50-nm-thick SiN film with an average temperature of 500 K. For the shortest film, T approaches to the linear profile predicted by Fourier's heat diffusion equation without heat sources, as a result of the ballistic propagation of polaritons ($\Lambda \gg l = 1$ mm) with very weak energy absorption and negligible energy exchange with phonons. The polariton heat generation rate is thus small, and their contribution to the heat flux is pretty much independent of position, as shown in Fig. 5.9b. For longer nanofilms with a length ($l = 5$ and 10 mm) comparable to the polariton propagation length, on the other hand, the quasiballistic propagation of

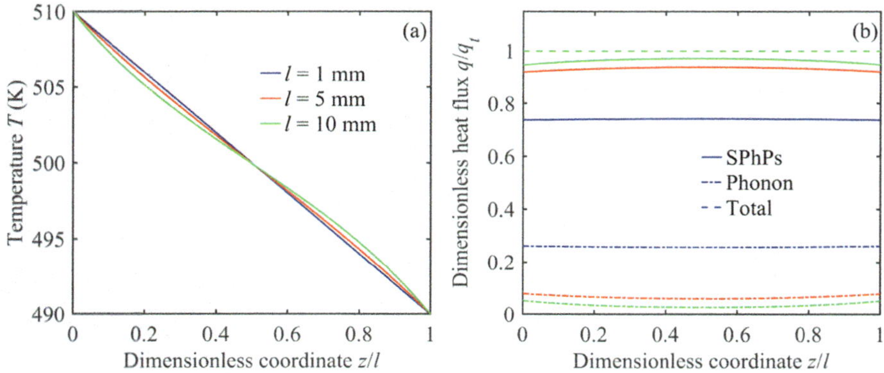

Fig. 5.9 Spatial distribution of the (**a**) temperature and (**b**) heat fluxes in a 50-nm-thick SiN film at 500 K. Calculations were done for three film lengths of practical interest

polaritons fosters their energy exchange with phonons, which generates a nonlinear temperature profile similar to that predicted by the heat diffusion equation with a heat sink for $z/l \leq 0.5$ and a heat source for $z/l \geq 0.5$, as seen in Fig. 5.9a. The apparent heat sink and heat source terms in the nanofilm arise from the predominant emission and adsorption of polaritons near its hot and cold sides, respectively. Therefore, the polariton heat flux increases with position until $z/l = 0.5$ and decreases afterward, while the phonon counterpart shows the opposite trend, as established by the principle of energy conservation. The increase of the polariton heat flux with the nanofilm length provides a pathway to enhance the in-plane heat transport along polar nanofilms by means of the combined dynamics of polaritons and phonons. Furthermore, as the nonlinearity of the temperature profile represents the fingerprints of polaritons, its experimental observation could provide a conclusive and easy way to detect the polariton heat transport.

5.2 Polariton Thermal Conductivity of Cylindrical Structures

Here we consider a polar nanowire supporting the simultaneous propagation of SPhPs and phonons due to the temperature difference $T_h - T_c$ set by two thermal baths, as shown in Fig. 5.10. The resulting steady-state heat conduction along the z axis is thus driven by the heat fluxes generated by these two energy carriers, and therefore the temperature profile is given by the conservation of energy in (5.1). Given that the total heat flux $q_t = q(\text{phonon})+q(\text{SPhP})$ is a constant independent of position z, the derivative of (5.1) with respect to z yields the following heat diffusion equation:

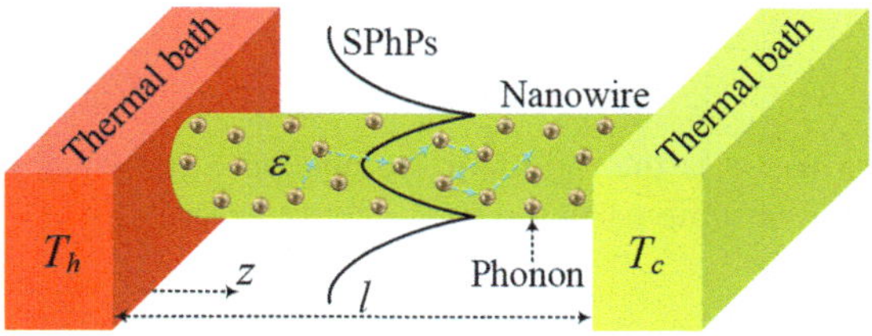

Fig. 5.10 Scheme of a nanowire supporting the propagation of both SPhPs and phonons due to the temperature difference $T_h - T_c$ imposed by two thermal baths

$$k_{ph}\frac{\partial^2 T}{\partial z^2} + S(z) = 0, \tag{5.38}$$

with $S(z) = -\partial q/\partial z$ being an effective heat source term that stands for the coupling between SPhPs and phonons. Physically, $S(z) > 0$ ($S(z) < 0$) represents the heat source (sink) due to the thermal absorption (emission) of SPhPs at the nanowire surface. As in the case of nanofilms, the SPhP heat flux q can be determined by means of the bosonic BTE under the relaxation time approximation. For the 1D propagation of SPhPs along the nanowire shown in Fig. 5.10, the steady-state BTE can conveniently be split into two components, as follows:

$$f^+ + \Lambda\frac{\partial f^+}{\partial z} = f_0, \tag{5.39}$$

$$f^- - \Lambda\frac{\partial f^-}{\partial z} = f_0, \tag{5.40}$$

where f is the SPhP distribution function, its equilibrium counterpart f_0 is the Bose–Einstein distribution function, and the superscript "$+(-)$" stands for the SPhP propagation along $+z(-z)$ direction a distance Λ. The analytical solutions of (5.39) and (5.40) for the distributions f^+ and f^- of the SPhPs leaving the surfaces $z = 0$ and $z = l$ are given by

$$f^+(\xi) = f_0(0)e^{-\xi} + \int_0^{\xi} f_0(\xi')e^{-(\xi'-\xi)}d\xi', \tag{5.41}$$

$$f^-(\xi) = f_0(\lambda)e^{-(\lambda-\xi)} + \int_{\xi}^{\lambda} f_0(\xi')e^{-(\xi-\xi')}d\xi', \tag{5.42}$$

where $\xi = z/\Lambda$, $\lambda = l/\Lambda$. In writing (5.41) and (5.42), we used the boundary conditions: $f^+(0) = f_0(0) \equiv f_0(T_h)$ and $f^-(\lambda) = f_0(\lambda) \equiv f_0(T_c)$ established by the thermal equilibrium of the external surfaces $\xi = 0$ and $\xi = \lambda$ set at the temperatures T_h and T_c, respectively. The correlation between $f^\pm$ and the SPhP heat flux is given by the definition of q: energy (E) per unit time (t) and unit cross-section, which in a differential form reads

$$dq = \frac{1}{\pi a^2}\frac{dE}{dt}, \tag{5.43}$$

where a is the radius of the nanowire shown in Fig. 5.10. Considering the SPhPs like bosonic particles, their energy is given by

$$dE = \hbar\omega f dN, \tag{5.44}$$

where $dN = dzdp/h$ is the differential of SPhP states, with $p = \hbar\beta_R$ being the SPhP momentum. Since Planck's constant $h = 2\pi\hbar$, $dN = dzd\beta_R/2\pi$ is determined by the product of the differentials of position and axial wave vector β_R. Therefore, the combination of (5.43) and (5.44) yields

$$dq = \frac{1}{2(\pi a)^2}\hbar\omega f d\omega, \tag{5.45}$$

where we have used definition of the group velocity $V = dz/dt = d\omega/d\beta_R$. The net SPhP heat flux q is thus given by the integration of (5.45) for all frequencies supporting the propagation of SPhPs in all polarizations p, as follows:

$$q = \frac{1}{2(\pi a)^2}\sum_p \int_0^\infty \hbar\omega\left(f^+ - f^-\right)d\omega. \tag{5.46}$$

After inserting (5.41) and (5.42) into (5.46), one obtains

$$q = \frac{1}{2(\pi a)^2}\sum_p \int_0^\infty \hbar\omega\left(f_0(0)e^{-\xi} - f_0(\lambda)e^{-(\lambda-\xi)} - \frac{d}{d\xi}\int_0^\lambda f_0(\xi')e^{-|\xi'-\xi|}d\xi'\right)d\omega. \tag{5.47}$$

For simplicity, (5.47) can be rewritten in terms of the normalized equilibrium intensity $U(\xi) = [f_0(\xi) - f_0(\lambda)]/[(f_0(0) - f_0(\lambda)]$, as follows:

$$q = \frac{1}{2(\pi a)^2}\sum_p \int_0^\infty \hbar\omega\left[f_0(0) - f_0(\lambda)\right]\left(e^{-\xi} - \frac{d}{d\xi}\int_0^\lambda U(\xi')e^{-|\xi'-\xi|}d\xi'\right)d\omega. \tag{5.48}$$

The SPhP heat flux q thus results from the equilibrium distribution difference $f_0(0)-f_0(\lambda) = f_0(T_h)-f_0(T_c)$ driven by the temperature difference $\Delta T = T_h-T_c$, as expected. Assuming that the nanowire undergoes small temperature gradients ($\Delta T \ll (T_h + T_c)/2 = T$), the Taylor series expansion of $f_0(T \pm \Delta T) = f_0(T) \pm \Delta T\partial f_0/\partial T + \ldots$ establishes that the first-order approximation of the difference $f_0(T_h) - f_0(T_c) \approx \Delta T\partial f_0/\partial T$ increases linearly with ΔT. Under this condition, the temperature profile exhibits a nearly linear dependence on position, and therefore f_0 can be approximated by $f_0(\xi) \approx \alpha(\beta - \xi)$, with α and β being two parameters independent of position. For this approximation, $U(\xi) = 1 - \xi/\lambda$ and (5.48) takes the form

$$q(\xi) = \frac{\Delta T}{2(\pi a)^2} \sum_p \int_0^\infty \hbar\omega \frac{\partial f_0}{\partial T} \left(\frac{2 - e^{-\xi} - e^{-(\lambda-\xi)}}{\lambda} \right) d\omega. \tag{5.49}$$

Note that the SPhP heat flux takes the same value ($q(\xi) = q(\lambda - \xi)$) at two equidistant positions from the external nanowire surfaces ($\xi = 0; \lambda$), such that its maximum appears at the middle of the nanowire ($\xi = \lambda/2$). As in this case of flat structures studied in Sect. 5.1, this position dependence of q arises from its nonlocal dependence on temperature, as established by (5.47). The integration of (5.1) for the SPhP heat flux in (5.49) and the boundary conditions $T(z = 0) = T_h$ and $T(z = l) = T_c$ yields the following total heat flux q_t and temperature profile:

$$q_t = \left(k + k_{ph}\right) \frac{\Delta T}{l}, \tag{5.50}$$

$$T(z) = T_h - \Delta T \frac{z}{l} + \frac{\Delta T}{2k_{ph}} \int_0^\infty k_\omega \left[\frac{\psi(z/\Lambda) + (2z/l - 1)\psi(0)}{\lambda - \psi(0)} \right] d\omega, \tag{5.51}$$

where $\psi(\xi) = e^{-\xi} - e^{-(\lambda-\xi)}$ and the spectral polariton thermal conductivity k_ω is defined in terms of its integrated counterpart $k = \int k_\omega d\omega$ given by

$$k = \frac{1}{(\pi a)^2} \sum_p \int_0^\infty \left(1 - \frac{\psi(0)}{\lambda}\right) \hbar\omega\Lambda \frac{\partial f_0}{\partial T} d\omega. \tag{5.52}$$

As the heat transport in a polar nanowire is driven by both phonons and SPhPs, the total heat flux is determined by the sum of thermal conductivities related to these two energy carriers, as established by (5.50). Therefore, the usual reduction of k_{ph} could be offset by the increase of the polariton thermal conductivity k, as the nanowire radius a scales down. In the diffusive refine ($\lambda = l/\Lambda \gg 1$), the ratio $\psi(0)/\lambda \to 0$ and k becomes independent of the nanowire length l. In the ballistic regime ($\lambda \ll 1$), on the other hand, $1 - \psi(0)/\lambda \approx \lambda/2$ and k gets independent of the propagation length Λ. In any case, the finite values of k drive the deviation of the temperature profile from its usual linear dependence on position, as determined by (5.51). Interestingly, the SPhP contribution to the temperature profile disappears at the middle of the nanowire ($z = l/2$) due to the symmetry of the SPhP heat flux around this position. As a result of this symmetry, the sum of temperatures at two equidistant points from the external nanowire surfaces is an invariant of heat conduction given by $T(z) + T(l - z) = T_h - T_c$.

In terms of the SPhP thermal conductance $G = \pi a^2 k/l$ of the nanowire, (5.52) takes the form of the following 1D Landauer formula [13, 14]:

$$G = \frac{1}{2\pi} \sum_p \int_0^\infty \hbar\omega \frac{\partial f_0}{\partial T} \tau(\omega) d\omega, \tag{5.53}$$

$$\tau = \frac{2}{\lambda}\left(1 - \frac{\psi(0)}{\lambda}\right), \tag{5.54}$$

where τ is the transmission probability of SPhPs to travel from one thermal bath to the other through the nanowire shown in Fig. 5.10). For long nanowires ($\lambda \gg 1$), the transmission probability goes to zero ($\tau \to 0$), and therefore G vanishes. By contrast, for short nanowires ($\lambda \ll 1$), $\tau \approx 1$, and (5.53) reduces to $G = NG_0$, where N is the number of SPhP modes with ballistic propagation along the nanowire and G_0 is the following quantum of thermal conductance:

$$G_0 = \frac{1}{2\pi} \int_0^\infty \hbar\omega \frac{\partial f_0}{\partial T} d\omega = \frac{k_B^2 T}{h} \int_0^\infty \frac{x^2 e^x}{(e^x - 1)^2} dx = \frac{\pi^2 k_B^2 T}{3h}. \tag{5.55}$$

The polariton contribution to the thermal conductance of a nanowire allowing the complete transmission of polaritons is thus independent of the material properties, depends on the universal constants k_B and h, and increases linearly with temperature. Since (5.55) was obtained by integrating over all frequencies, this quantization of the polariton thermal conductance holds for a SPhP mode without forbidden bands of frequency. The comparison of (5.21) and (5.55) indicates that the temperature dependence of the quantum of thermal conductance of polaritons is given by $G_0 \propto T^n$, where n is determined by the dominant dimensionality of nanowires ($n = 1$) and nanofilms ($n = 2$). Outside the ballistic regime, the 1D SPhP propagation along nanowires yields a higher transmission probability than its 2D counterpart along nanofilms, as shown in Fig. 5.11. The transmissivity difference takes its maximum $\tau(\text{1D}) - \tau(\text{2D}) \approx 8\%$ for a length $l = 1.8\Lambda$, which defines a diffusive-ballistic propagation. The 1D channel of nanowires is thus slightly better than the 2D one of nanofilms for the transmission of the polariton thermal energy.

We now quantify and analyze the propagation parameters and heat transport of SPhPs propagating along a nanowire of SiN. This polar material is able to support the propagation of SPhPs in a wide frequency range and therefore is considered to be a good polariton conductor. The wave vector β_R and propagation length $\Lambda = [2\text{Im}(\beta)]^{-1}$ of SPhPs propagating along the surface of a SiN nanowire suspended in vacuum are shown in Fig. 5.12, as functions of frequency. Note that $\text{Re}(\beta)$ generally increases with frequency, through values higher than those of the light line ($k_0 = \omega/c$). The deviation from this monotonic behavior shows up around 31.9 THz, which is pretty close to the frequency (31.8 THz) where the real part of the SiN permittivity changes its sign. The relatively weak absorption of the thin nanowire ($R = 50$ nm) enables SPhPs to propagate distances as long as 2.7 cm at high frequency (75 THz). The minimum of Λ is caused by the maximum

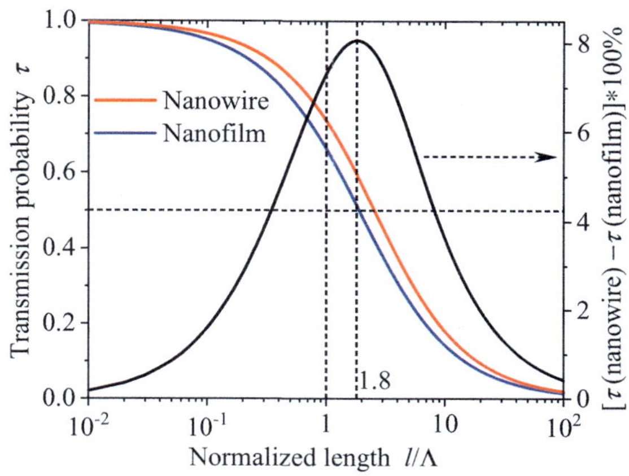

Fig. 5.11 Probability that polaritons transmit along a nanowire and nanofilm. Calculations were done with (5.37) and (5.54)

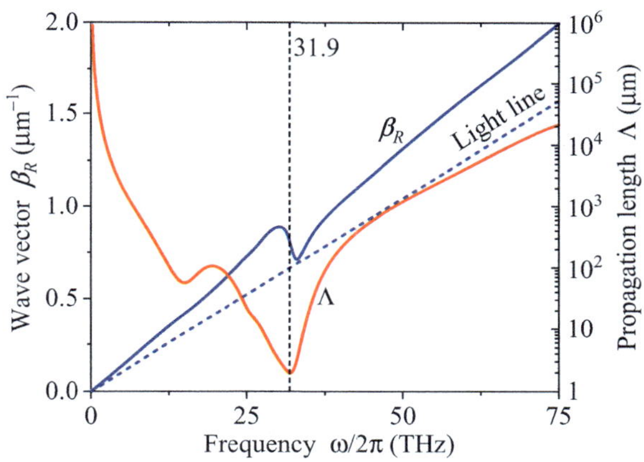

Fig. 5.12 Frequency spectrum of the wave vector and propagation length of SPhPs propagating along a SiN nanowire suspended in vacuum. The blue dashed line stands for the wave vector of light in vacuum. Calculations were done for the long-range mode predicted by the dispersion relation in (4.164), $R = 50$ nm, and $\varepsilon_2 = 1$

energy absorption and negligible value of ε_R at 31.9 THz. In addition, the fact that $\Lambda > 1\ \mu$m indicates that SPhPs propagate ballistically along a SiN nanowire shorter than $l < 1\ \mu$m, and therefore their contribution to the heat transport could be driven by the quantum of thermal conductance G_0 of a nanowire.

Figure 5.13 shows the frequency spectrum of the SPhP transmission probability τ for four lengths of a SiN nanowire. Note that shorter nanowires exhibit a higher transmissivity, whose lowest value at 32.1 THz is correlated to the minimum value of the SPhP propagation length shown in Fig. 5.12. By contrast, for other sufficiently low and high frequencies, τ tends to unity as a result of the long propagation lengths of SPhPs. These lowest and highest values of τ drive the behavior of the thermal conductivity spectrum, which takes higher values for shorter nanowires, as shown in Fig. 5.14a. At high enough frequencies, this spectrum becomes independent of the nanowire length and decays exponentially due to the insufficient thermal energy required to excite them. At very low frequencies, on the other hand, the thermal conductance spectrum takes its highest values due to the high transmissivity

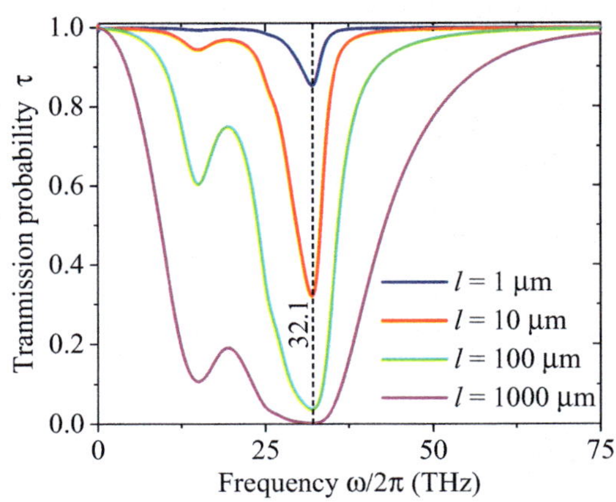

Fig. 5.13 Frequency spectrum of the transmission probability of SPhPs propagating along a SiN nanowire of different lengths

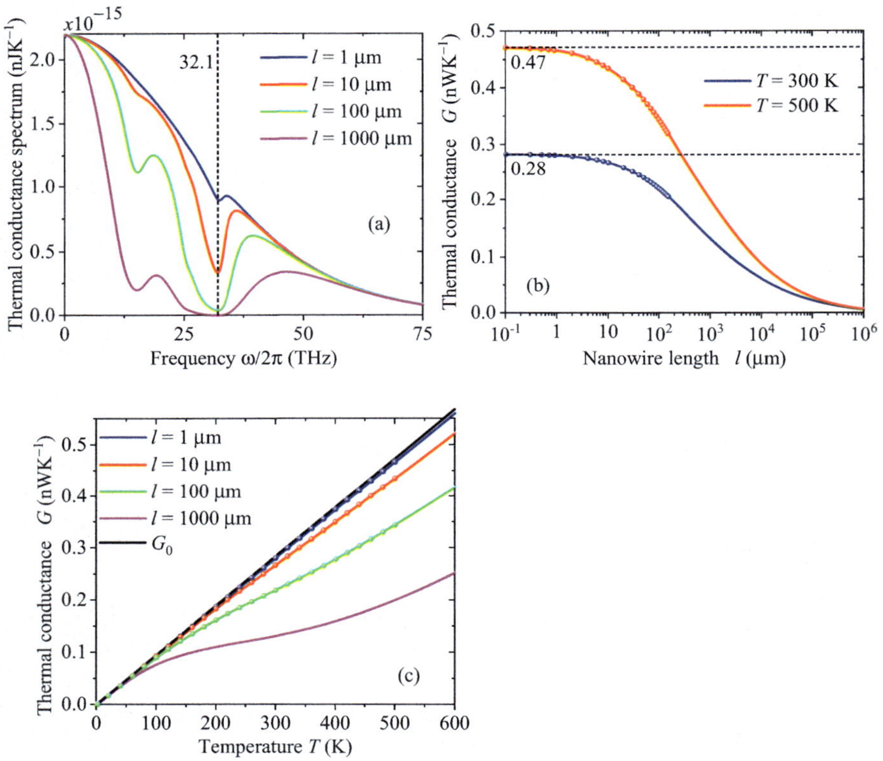

Fig. 5.14 (**a**) SPhP thermal conductance spectrum along with its integrated counterpart as a function of the (**b**) length and (**c**) temperature of a SiN nanowire suspended in vacuum with a radius $R = 50\,\text{nm}$. The solid lines represent the predictions of (5.53), and the dots stand for the numerical solution of the BTE [14]. Calculations in (**a**) were carried out for $T = 300\,\text{K}$

of SPhPs. The integration of the spectra in Fig. 5.14a yields the SPhP thermal conductance G shown in Fig. 5.14b, c, as a function of the nanowire length and temperature, respectively. As a result of the high transmissivity, shorter nanowires exhibit a higher G, whose values increase with temperature. The very good agreement between the analytical (solid lines) and numerical (dots) calculations, for the three temperatures and lengths, confirms the high accuracy of (5.53) for predicting the thermal conductance of SPhPs. More importantly, the upper bound of G, in the ballistic regime ($l < 1$ μm), is well confirmed by the analytical and numerical solutions, and its values coincide with the quantum of thermal conductance G_0. Even though the transmissivity of a 1 μm-long SiN nanowire is not unity over the full frequency spectrum (see Fig. 5.13), the corresponding SPhP thermal conductance is pretty much equal to G_0. The quantization of the SPhP thermal conductance is thus expected to be observed in SiN nanowires with a length comparable to or shorter than 1 μm.

The temperature and heat flux profiles along a SiN nanowire supporting the simultaneous propagation of SPhPs and phonons are, respectively, shown in Fig. 5.15a, b, for three nanowire lengths. For the shortest nanowire with a length ($l = 10$ μm) much smaller than the propagation length of most SPhPs, SPhPs propagate ballistically with weak adsorption and low energy exchange with phonons. The SPhP heat generation inside the nanowire is therefore small, and T exhibits the typical linear behavior predicted by the heat diffusion, without a heat source term. This fact is confirmed by the relatively low SPhP heat flux, which is nearly independent of position due to the decoupling of phonons and SPhPs, as shown in Fig. 5.15b. On the other hand, for longer nanowires with a length ($l = 80$ and 150 μm) comparable to the propagation length of some SPhPs, the quasiballistic propagation of SPhPs fosters their energy exchange with phonons and generates a

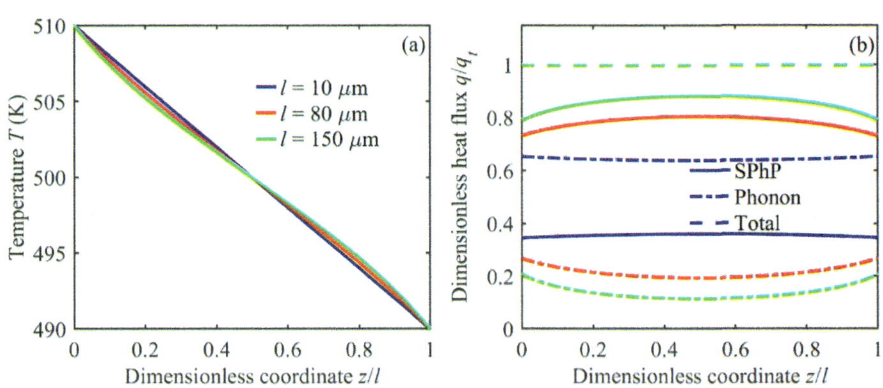

Fig. 5.15 (**a**) Temperature and (**b**) heat flux profiles in a SiN nanowire supporting the simultaneous propagation of SPhPs and phonons. The solid and dashed-dot lines in (**b**) represent the respective SPhP and phonon heat fluxes, whereas the dashed one stands for the total heat flux q_t. Calculations were carried out for a nanowire with radius $R = 50$ nm, three lengths l, and a typical phonon thermal conductivity of $k_{ph} = 1\,\mathrm{Wm^{-1}K^{-1}}$

nonlinear temperature profile similar to that predicted by the heat diffusion equation with a heat sink for $z/l \leq 0.5$ and a heat source for $z/l \geq 0.5$. The apparent heat sink and heat source terms in the nanowire arise from the predominant emission and absorption of SPhPs near its hot and cold sides, respectively. Therefore, the SPhP heat flux increases with position until $z/l = 0.5$ and decreases afterward, while the phonon counterpart shows the opposite trend, as established by the principle of energy conservation in (5.1). The increase of the SPhP heat flux with the nanowire length provides a pathway to enhance the heat transport along polar nanowires by means of the coupling of SPhPs and phonons, as is the case in polar nanofilms studied in Sect. 5.1. Furthermore, given that the nonlinear temperature profile represents the fingerprints of SPhPs, its experimental observation provides an intuitive and conclusive way to detect the SPhP heat transport. For instance, for a 150-μm-long SiN nanowire, the largest temperature deviation from the linear profile is about 1 K (see Fig. 5.15a), which is measurable by infrared thermometers.

References

1. Chen, G.: Nanoscale energy transport and conversion. Oxford University Press, New York (2005)
2. Chen, S.A., Narayanaswamy, A., Chen, G.: Surface phonon-polariton mediated thermal conductivity enhancement of amorphous thin films. Phys. Rev. B **72**, 155435 (2005)
3. Guo, Y., Tachikawa, S., Volz, S., Nomura, M., Ordonez-Miranda, J.: Quantum of thermal conductance of nanofilms due to surface-phonon polaritons. Phys. Rev. B **104**, L201407 (2021)
4. Mingo, N., Broido, D.A.: Carbon nanotube ballistic thermal conductance and its limits. Phys. Rev. Lett. **95**, 096105 (2005)
5. Saito, K., Nakamura, J., Natori, A.: Ballistic thermal conductance of a graphene sheet. Phys. Rev. B **76**, 115409 (2007)
6. Bae, M.-H., Li, Z., Aksamija, Z., Martin, P.N., Xiong, F., Ong, Z.-Y., Knezevic, I., Pop, E.: Ballistic to diffusive crossover of heat flow in graphene ribbons. Nat. Commun. **4**, 1734 (2013)
7. Tachikawa, S., Ordonez-Miranda, J., Jalabert, L., Wu, Y.,Anufriev, R., Guo, Y., Kim, B., Fujita, H., Volz, S., Nomura, M.: Enhanced far-field thermal radiation through a polaritonic waveguide. Phys. Rev. Lett. **132**, 186904 (2024)
8. Kourganoff, V.: Basic Methods in Transfer Problems. Dover, New York (1963)
9. Imry, Y., Landauer, R..: Conductance viewed as transmission. Rev. Mod. Phys. **71**, S306–S312 (1999)
10. Wu, Y., Ordonez-Miranda, J., Gluchko, S., Anufriev, R., De Sousa Meneses, D., Del Campo, L., Volz, S., Nomura, M.: Enhanced thermal conduction by surface phonon-polaritons. Sci. Adv. **6**, eabb4461 (2020)
11. Palik, E. D.: Handbook of Optical Constants of Solids. Academic Press, Orlando (1985)
12. Ordonez-Miranda, J., Tranchant, L., Tokunaga, T., Kim, B., Palpant, B., Chalopin, Y., Antoni, T., Volz, S.: Anomalous thermal conductivity by surface phonon-polaritons of polar nanofilms due to their asymmetric surrounding media. J. Appl. Phys. **113**, 084311 (2013)
13. Ordonez-Miranda, J., Tranchant, L., Kim, B., Chalopin, Y., Antoni, T., Volz, S.: Quantized thermal conductance of nanowires at room temperature due to zenneck surface-phonon polaritons. Phys. Rev. Lett. **112**, 055901 (2014)
14. Guo, Y., Nomura, M., Volz, S., Ordonez-Miranda, J.: Heat transport driven by the coupling of polaritons and phonons in a polar nanowire. Energies **14**, 5110 (2021)

Chapter 6
Polariton Heat Transport in Metals

6.1 Plasmon Heat Transport Along Metallic Nanofilms

In Chap. 5 we saw that suspended polar nanofilms are symmetrical structures able to support the long-range propagation of SPhPs in a broad range of frequencies. By contrast, for supported structures, such as a nanofilm deposited on a substrate, the propagation distance and frequency spectrum of SPhPs are significantly reduced due to the mismatch of the emissivities of its surrounding media and the existence of the SPhP resonance of polar materials [Re(permittivity) $<$ 0] in a limited frequency interval. These two reductions significantly suppress the SPhP contribution to the thermal conductivity of supported polar nanofilms [1], even though they provide better conditions than suspended ones for practical use. In general, suspended nanofilms and symmetric structures favor the propagation and in-plane heat transport of SPhPs.

By analogy with polar dielectrics supporting the propagation of SPhPs, metals support the propagation of SPPs resulting from the hybridization of photons and free electrons propagating near the metal interfaces. The abundance of these free electrons in metals leads to a high plasma frequency ω_p that determines the upper limit $(\omega_p/\sqrt{2})$ of the frequency spectrum supporting the propagation of SPPs [2]. In contrast to polar dielectrics, the high ω_p of metals (i.e., $\omega_p/2\pi = 2196.34$ THz for gold) and the relatively large negative values of the real part of their permittivity for frequencies smaller than ω_p enable the existence and propagation of SPPs in a broad range of frequencies. This wide spectrum of SPP propagation is typically excited with thermal sources and can significantly enhance the near-field radiation (cross-plane propagation) and heat conduction (in-plane propagation) in metallic nanofilms. In particular, the fact that SPPs can propagate distances of a few centimeters along a nanofilm, even in the presence of a substrate, makes of them powerful heat carriers. As metallic nanofilms with these lateral dimensions are easy

S. Volz, J. Ordonez-Miranda, *Heat Transport Driven by Surface Electromagnetic Waves*, Mechanical Engineering Series, https://doi.org/10.1007/978-3-031-66024-5_6

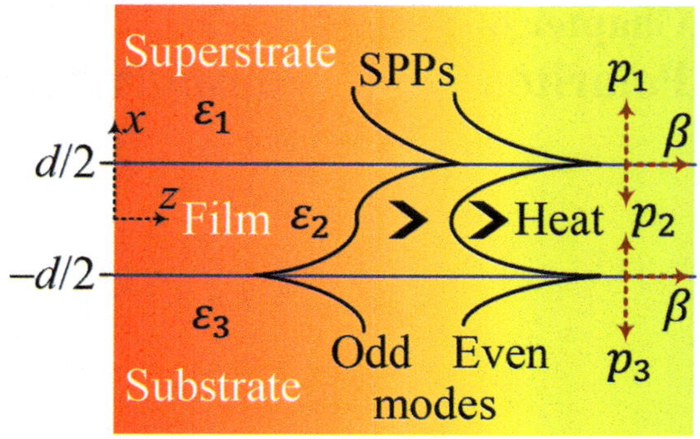

Fig. 6.1 Scheme of a metallic film supporting the propagation of SPPs along its interfaces with the substrate and superstrate of relative permittivities ε_1 and ε_3, respectively. The film has a relative permittivity ε_2 and thickness d. The SPPs and their heat flux propagate from the red zone (high temperature) to the yellow one (low temperature)

to fabricate by depositing them on a substrate, they provide a suitable medium to capitalize on the full propagation potential of SPPs.

To determine the SPP contribution to the thermal conductance and thermal conductivity of a metallic nanofilm, we consider the propagation of SPPs along its interfaces with the substrate and superstrate, as shown in Fig. 6.1. These top and bottom materials have relative permittivities ε_1 and ε_3, respectively; and they are separated by the nanofilm of relative permittivity ε_2 and thickness d. Assuming that the surface $x = 0$ is uniformly heated up with a thermal bath, the nanofilm free electrons oscillate and emit an electromagnetic field. These oscillations induce the excitation of neighboring electrons, which keeps the field propagation along the film interfaces mainly. This thermally excited field represents an SPP able to propagate in a broad range of frequencies. Under this condition, the heat propagates along the x axis via the combined dynamics of electrons and SPPs. Given that SPPs, as SPhPs, are bosons [2], the SPP thermal conductance G of the nanofilm is given (5.36). Taking into account that the values of G increase with the propagation length $\Lambda = [2\mathrm{Im}(\beta)]^{-1}$, we focus on the in-plane wave vector β of the long-range modes predicted for the TM polarization. According to (4.2) and (4.3), the in-plane field components for the TM waves propagating inside and outside the nanofilm shown in Fig. 6.1 can be written as follows:

$$E_z^{(1)} = A_1 e^{-p_1(x-d/2)}, \quad H_y^{(1)} = \frac{i\omega\varepsilon_1\varepsilon_0}{p_1} E_z^{(1)}, \tag{6.1}$$

$$E_z^{(2)} = A \sinh(p_2 x) + B \cosh(p_2 x), \quad H_y^{(2)} = -\frac{i\omega\varepsilon_2\varepsilon_0}{p_2^2} \frac{\partial E_z^{(2)}}{\partial x}, \tag{6.2}$$

$$E_z^{(3)} = A_3 e^{p_3(x+d/2)}, \quad H_y^{(3)} = -\frac{i\omega\varepsilon_3\varepsilon_0}{p_3} E_z^{(3)}, \tag{6.3}$$

where the cross-plane wave vectors p_n are given by $p_n^2 = \beta^2 - \varepsilon_n k_0^2$. The continuity of these tangential field components at the interfaces $x = \pm d/2$ yields

$$A_1 = A\sinh(\alpha) + B\cosh(\alpha) = -S_{21}\left[A\cosh(\alpha) + B\sinh(\alpha)\right], \tag{6.4}$$

$$A_3 = -A\sinh(\alpha) + B\cosh(\alpha) = S_{23}\left[A\cosh(\alpha) - B\sinh(\alpha)\right], \tag{6.5}$$

where $S_{nm} = S_n/S_m$, $S_n = \varepsilon_n/p_n$, and $\alpha = p_2 d/2$. The parameters A and B are therefore determined by the following system of equations:

$$\left[S_{21} + \tanh(\alpha)\right]A = -\left[1 + S_{21}\tanh(\alpha)\right]B, \tag{6.6}$$

$$\left[S_{23} + \tanh(\alpha)\right]A = \left[1 + S_{23}\tanh(\alpha)\right]B. \tag{6.7}$$

As we are solving the Maxwell's equations for a non-null electromagnetic field inside the film, A and B cannot be zero simultaneously. Under this physical condition, (6.6) and (6.7) establish that these two parameters are not zero, and hence their solution for β is given by

$$\frac{S_{21} + \tanh(\alpha)}{S_{23} + \tanh(\alpha)} = -\frac{1 + S_{21}\tanh(\alpha)}{1 + S_{23}\tanh(\alpha)}. \tag{6.8}$$

By rewriting (6.8) as a polynomial on $\tanh(\alpha)$, we obtain the following SPP dispersion relation:

$$\frac{p_2}{\varepsilon_2}\left(\frac{p_1}{\varepsilon_1} + \frac{p_3}{\varepsilon_3}\right) + \left(\frac{p_2^2}{\varepsilon_2^2} + \frac{p_1}{\varepsilon_1}\frac{p_3}{\varepsilon_3}\right)\tanh\left(p_2 d\right) = 0, \tag{6.9}$$

where we have used the identity $\tanh(2\alpha) = 2\tanh(\alpha)/[1 + \tanh^2(\alpha)]$. As (6.9) was derived by matching the tangential components of the electric and magnetic fields at the film interfaces $x = \pm d/2$, it is valid for a nanofilm of arbitrary length. When the nanofilm is surrounded by a single medium ($\varepsilon_1 = \varepsilon_3$), (6.9) splits into the dispersion relations of the symmetric and antisymmetric TM modes in (4.61) and (4.84), as expected. For a very thin film ($d \to 0$), (6.9) becomes independent of the film parameters and reduces to the well-known dispersion relation $p_1/\varepsilon_1 + p_3/\varepsilon_3 = 0$ ($\beta = \sqrt{\varepsilon_1\varepsilon_3/(\varepsilon_1 + \varepsilon_3)}$) of a single interface between the substrate and superstrate. On the other hand, for a very thick film ($|p_2|\, d \gg 1$), the SPPs propagating along the two film interfaces decouple and propagate with a wave vector $\beta_n = \sqrt{\varepsilon_n\varepsilon_2/(\varepsilon_n + \varepsilon_2)}$ for $n = 1$ and 3. None of these two limiting cases thus allows us to capitalize on the SPP coupling across the film to enhance its SPP heat transport. The effect of the film thickness on G is therefore expected to appear for intermediate thicknesses. As the surface confinement of SPPs in medium $n = 1, 2, 3$ is driven by $\mathrm{Re}\,(p_n)$, the solution of (6.9) for the SPP wave vectors $\beta = k_0\sqrt{\varepsilon}$ and $p_n = k_0\sqrt{\varepsilon - \varepsilon_n}$ is generally determined by effective permittivity ε changing between ε_1 and ε_3. This constraint on the possible values of ε indicates that when the permittivity amplitude of the metallic film is much greater than the corresponding ones of its surrounding media ($|\varepsilon_2| \gg |\varepsilon_1|$ and $|\varepsilon_3|$), the transverse wave vector $p_2 = k_0\sqrt{\varepsilon - \varepsilon_2} \approx k_0\sqrt{-\varepsilon_2}$ becomes independent of the film thickness and permittivities of the surrounding media. Under this key condition, (6.9) takes the form

$$\frac{\sqrt{\rho - \varepsilon_{12}}}{\varepsilon_{12}} + \frac{\sqrt{\rho - \varepsilon_{32}}}{\varepsilon_{32}} + \left(\frac{\sqrt{\rho - \varepsilon_{12}}}{\varepsilon_{12}}\frac{\sqrt{\rho - \varepsilon_{32}}}{\varepsilon_{32}} - 1\right)\gamma = 0, \tag{6.10}$$

where $\rho = \varepsilon/\varepsilon_2$, $\varepsilon_{nm} = \varepsilon_n/\varepsilon_m$, $\gamma = \tan\left(\lambda\sqrt{\varepsilon_2}\right)$, and $\lambda = k_0 d$. To better understand the solutions of (6.10), the following limiting cases are considered:

6.1.1 *Thick Film Limit: $d > 4\delta_2$*

Taking into account the identity $\tan(x + iy) = [\tan(x) + i\tanh(y)]/[1 - i\tan(x)\tanh(y)] = i$ for $\tanh(y) = 1$, which is well satisfied for $y > 2$, (6.10) can be solved with $\gamma = i$ for $\lambda\mathrm{Im}(\sqrt{\epsilon_2}) > 2$. In this case, the two solutions of (6.10) are given by $\sqrt{\rho - \epsilon_{n2}} = i\epsilon_{n2}$, which yields the following effective permittivity:

$$\varepsilon = \varepsilon_n\left(1 - \frac{\varepsilon_n}{\varepsilon_2}\right), \tag{6.11}$$

for $n = 1$ and 3. The first solution ($n = 1$) is related to the SPP propagation along the superstrate–film interface, while the second one ($n = 3$) drives the propagation along the film–substrate interface. As each of these two solutions depends on the permittivity of a single surrounding material, the SPP propagation along an interface is not affected by the one along the other interface. This SPP decoupling arises due to the fact that the film thickness $d = \lambda/k_0 > 2\left[k_0\mathrm{Im}(\sqrt{\epsilon_2})\right]^{-1} = 2\left[k_0\mathrm{Re}(\sqrt{-\epsilon_2})\right]^{-1} = 4\delta_2$ is greater than four times the SPP penetration depth $\delta_2 = [2\mathrm{Re}\,(p_2)]^{-1}$ inside the film. In terms of the real and imaginary parts of the film permittivity $\varepsilon_2 = \varepsilon_R + i\varepsilon_I$, this later condition reads

$$\varepsilon_I^2(\omega) > \left(\frac{4}{\lambda}\right)^2\left[\varepsilon_R(\omega) + \left(\frac{2}{\lambda}\right)^2\right]. \tag{6.12}$$

This parabolic inequality thus determines the range of frequencies in which (6.11) applies. This latter equation establishes that the cross-plane wave vectors $p_n = k_0\sqrt{\varepsilon - \varepsilon_n}$ are given by

$$\frac{p_1}{k_0\varepsilon_1} = \frac{p_3}{k_0\varepsilon_3} = \frac{1}{\sqrt{-\varepsilon_2}} = \frac{\sqrt{|\varepsilon_2| - \varepsilon_R} + i\sqrt{|\varepsilon_2| + \varepsilon_R}}{\sqrt{2}\,|\varepsilon_2|}, \tag{6.13}$$

$$\frac{p_2}{k_0} = \sqrt{-\varepsilon_2} = \frac{1}{\sqrt{2}}\left(\sqrt{|\varepsilon_2| - \varepsilon_R} - i\sqrt{|\varepsilon_2| + \varepsilon_R}\right). \tag{6.14}$$

Taking into account that the confinement of SPPs to both film interfaces increases with $\mathrm{Re}(p_n)$, (6.13) and (6.14) indicate that this confinement becomes stronger in metallic films with higher absorption ($\varepsilon_I \gg 1$) and/or $\varepsilon_R \ll 0$. If the substrate and superstrate are nonabsorbing media ($\mathrm{Im}(\varepsilon_1) = \mathrm{Im}(\varepsilon_3) = 0$), the real and imaginary parts of the in-plane wave vector $\beta = \beta_R + i\beta_I = k_0\sqrt{\varepsilon_n}\sqrt{1 - \varepsilon_n/\varepsilon_2} \approx k_0\sqrt{\varepsilon_n}\,(1 - \varepsilon_n/2\varepsilon_2)$, for $n = 1$ and 3, are given by

$$\beta_R = k_0\sqrt{\varepsilon_n}\left(1 - \frac{\varepsilon_n\varepsilon_R}{2\,|\varepsilon_2|^2}\right), \tag{6.15}$$

$$\beta_I = k_0\frac{\varepsilon_I\sqrt{\varepsilon_n}^3}{2\,|\varepsilon_2|^2}. \tag{6.16}$$

Equation (6.16) shows that the SPP propagation length $\Lambda = (2\beta_I)^{-1} \propto \varepsilon_I^{-1}\varepsilon_n^{-3/2}$ increases as the film absorption decreases ($\varepsilon_I \to 0$), and it takes longer values along the film interface with the medium of lower permittivity. Therefore, the spatial decay of SPPs along the film interfaces increases with ε_I. Further, according to (6.15), the SPPs propagate the distance Λ with a wave vector $\beta_R > 0$, for frequencies satisfying the condition $|\varepsilon_2|^2 > \varepsilon_n\varepsilon_R/2$, which yields

$$\left(\varepsilon_R - \frac{\varepsilon_n}{4}\right)^2 + \varepsilon_I^2 > \left(\frac{\varepsilon_n}{4}\right)^2. \tag{6.17}$$

In the plane $(\varepsilon_R, \varepsilon_I)$, the constraint in (6.17) establishes that the SPPs propagate outside the semicircle centered at $(\varepsilon_n/4, 0)$ with a radius $\varepsilon_n/4$, as $\varepsilon_I > 0$. Under the thick film limit ($d > 4\delta_2$), the existence and propagation of SPPs are thus determined by (6.12) and (6.17), respectively. Both constraints are satisfied by a relatively high imaginary part ε_I falling within the highlighted area shown in Fig. 6.2. Note that, for $\varepsilon_R < 0$, the values of ε_I allowing the SPP propagation are only determined by the parabolic constraint in (6.12). Considering that the permittivity of the metallic film is described by the Drude model, as is usually the

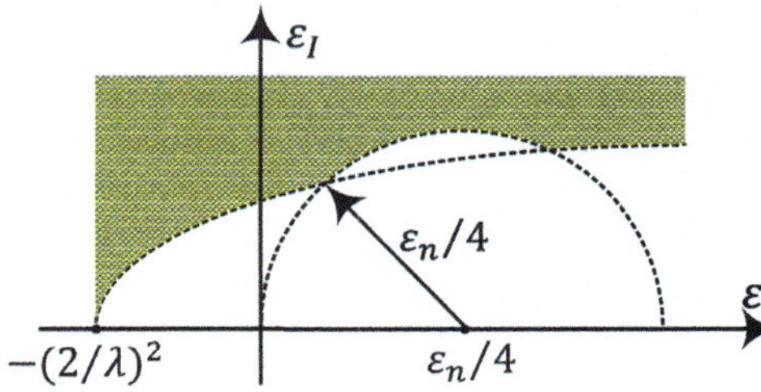

Fig. 6.2 Map of the region supporting the existence and propagation of SPPs along a thick film ($d > 4\delta_2$) with complex permittivity $\varepsilon_2 = \varepsilon_R + i\varepsilon_I$

case for a wide variety of pure metals, the frequency dependence of ε_2 is given by

$$\varepsilon_2(\omega) = 1 - \frac{\omega_p^2}{\omega^2 + i\Gamma\omega}, \tag{6.18}$$

where ω_p is the plasma frequency and Γ a damping factor driving the material energy absorption. After inserting (6.18) into (6.13)–(6.16), the SPP wave vectors reduce to

$$\frac{p_1}{k_0\varepsilon_1} = \frac{p_3}{k_0\varepsilon_3} = \frac{\omega}{\sqrt{2}\omega_p}\left(\sqrt{\chi+1} + i\sqrt{\chi-1}\right), \tag{6.19}$$

$$p_2 = \frac{\omega_p}{\sqrt{2}c}\frac{\sqrt{\chi+1} - i\sqrt{\chi-1}}{\chi}, \tag{6.20}$$

$$\beta_R = k_0\sqrt{\varepsilon_n}\left[1 + \frac{\varepsilon_n}{2}\left(\frac{\omega}{\omega_p}\right)^2\right], \tag{6.21}$$

$$\beta_I = \frac{\Gamma\sqrt{\varepsilon_n}^3}{2c}\left(\frac{\omega}{\omega_p}\right)^2, \tag{6.22}$$

where $\chi = \sqrt{1 + (\Gamma/\omega)^2}$ and the practical approximation $\omega \ll \omega_p$ was used due to the relatively high plasma frequency ($\omega_p/2\pi \gtrsim 100\,\text{THz}$) of usual metals [3]. Note that, for frequencies much lower than ω_p, but much higher than Γ ($\Gamma \ll \omega \ll \omega_p$), $\chi \approx 1$ and the SPP penetration depth $\delta_2 = [2\text{Re}(p_2)]^{-1} \approx 0.5c/\omega_p$ inside the metal film become independent of frequency, as established by (6.20). For a gold film ($\omega_p/2\pi = 2196.34\,\text{THz}$), for instance, this characteristic penetration depth $\delta_2 = 10.9\,\text{nm}$, which indicates that the film becomes thick for the SPP propagation when its thickness $d > 4\delta_2 = 43.6\,\text{nm}$. By contrast, the penetration depth within the surrounding media reduces to $\delta_n = [2\text{Re}(p_n)]^{-1} \approx c\omega_p/\varepsilon_n\omega^2$, which is inversely proportional to the square of frequency. Therefore, the confinement of SPPs outside and near the two film interfaces increases with the frequency ($\delta_n^{-1} \propto \omega^2$). This frequency dependence of δ_n is also exhibited by the propagation length $\Lambda = (2\beta_I)^{-1}$, which increases as the absorption parameter Γ decreases.

The explicit and simple frequency dependence of β_R in (6.21) allows to analytically determine the SPP thermal conductance of a metallic nanofilm supporting the ballistic propagation of SPPs ($\Lambda \gg l$). For these long-range SPPs, the transition probability becomes unity ($\tau = 1$), and (5.36) predicts the following contribution G_n of the modes $n = 1$ and 3 to the SPP thermal conductance $G = G_1 + G_3$

$$\frac{G_n}{a} = G_0\sqrt{\varepsilon_n}\left[1 + 10\frac{z(5)}{z(3)}\varepsilon_n\left(\frac{k_bT}{\hbar\omega_p}\right)^2\right], \tag{6.23}$$

where $G_0 = 12z(3)k_B^3T^2/ch^2$ is the quantum of thermal conductance of polaritons propagating along a nanofilm. The ballistic heat transport of SPPs therefore increases with temperature and the permittivity of the surrounding media, but it takes lower values along metallic films with higher plasma frequencies.

6.1.2 High Permittivity Limit: $|\varepsilon_2| \gg |\varepsilon_1|$ and $|\varepsilon_3|$

When the amplitude of the metallic film permittivity is much greater than the ones of the surrounding media, the solution of (6.10) can conveniently be obtained in terms of a series expansion on $\varepsilon_{n2} = \varepsilon_n/\varepsilon_2$, regardless of the thickness d. The square roots and symmetric dependence on ε_{12} and ε_{32} of (6.10) indicate that its solutions for $n = 1$ and $n = 3$ accept the following expansion:

$$\rho = \varepsilon_{n2} + (\alpha\varepsilon_{n2})^2 \left(1 + a_1\varepsilon_{n2} + a_2\varepsilon_{n2}^2 + \ldots\right). \tag{6.24}$$

For an approximation up to ε_{n2}^4 and $n = 1$, the combination of (6.10) and (6.24) yields the following expressions for the parameters α, a_1, and a_2:

$$\alpha = \gamma - \frac{1+\gamma^2}{\gamma - i\sqrt{\varepsilon_{32}}}, \tag{6.25}$$

$$a_1 = \frac{\sqrt{\varepsilon_{23}}\left(1+\gamma^2\right)}{\left(\gamma - i\sqrt{\varepsilon_{32}}\right)\left(\gamma\sqrt{\varepsilon_{32}} - i\right)}, \tag{6.26}$$

$$a_2 = -\frac{\sqrt{\varepsilon_{23}}^3\left(1+\gamma^2\right)\left(\gamma\sqrt{\varepsilon_{32}}A + iB\right)}{4\left(\gamma - i\sqrt{\varepsilon_{32}}\right)^3\left(\gamma\sqrt{\varepsilon_{32}} - i\right)^2}, \tag{6.27}$$

where $A = 3 - 11\varepsilon_{32} + 4\gamma^2\left(\varepsilon_{32}^2 - 1\right)$ and $B = 4\varepsilon_{32} - \gamma^2(3\varepsilon_{32}+1)(4\varepsilon_{32}-3)$. These three coefficients determine all wave vectors via the effective permittivity $\varepsilon = \varepsilon_2\rho$, which, according to (6.24), is given by

$$\varepsilon = \varepsilon_1\left[1 + \varepsilon_{12}\alpha^2\left(1 + a_1\varepsilon_{12} + a_2\varepsilon_{12}^2\right)\right]. \tag{6.28}$$

Note that ε is not only determined by the permittivities of both surrounding media but also by the film thickness d through the parameter γ. In the thick film limit ($d > 4\delta_2$), $\gamma = i = \alpha$, $a_1 = a_2 = 0$, and (6.28) reduces to (6.11), as expected. Equation (6.28) thus represents the solution of the SPP dispersion relation for an arbitrary film thickness, provided that $|\varepsilon_2| \gg |\varepsilon_1|$ and $|\varepsilon_3|$. This first solution ($n = 1$) for ε drives the propagation of SPPs along the interface with the superstrate and under the influence of the SPPs propagating along the other interface. The symmetry of (6.10) on ε_1 and ε_3 establishes that the second solution ($n = 3$) is related to the SPP propagation along the interface with the substrate and is given by (6.25)–(6.28) under the exchange of the subindexes $1 \leftrightarrow 3$.

The propagation and thermal properties of SPPs propagating along a gold nanofilm deposited on a silicon substrate ($\varepsilon_3 = 11.7$) and surrounded by air ($\varepsilon_1 = 1$) are now going to be determined numerically. These two surrounding media do not absorb energy in a large frequency range, and therefore they favor the propagation of SPPs over long distances. On the other hand, gold is a good SPP conductor due to its high plasma frequency ($\omega_p/2\pi = 2196.34$ THz) and relatively low absorption ($\Gamma/2\pi = 15.92$ THz), which enable the long-range propagation of SPPs in a wide frequency range. With these average values of ω_p and Γ, the Drude model describes well the gold permittivity [3].

The in-plane wave vector β_R and propagation length Λ of SPPs propagating along an gold nanofilm are respectively shown in Fig. 6.3a, b, for four representative film thicknesses. The mode $n = 1$ ($n = 3$) drives the SPP propagation along the air–film (film–substrate) interface. For both modes, β_R linearly increases with frequency through values pretty much equal to the wave vector ($\sqrt{\varepsilon_n}k_0$) of light in medium n, regardless of the film thickness, for the considered four values. This photon-like nature of SPPs indicates that their propagation along a film interface is nearly independent of the one along the other interface. For the mode propagating along the interface with Si, the surrounding medium with higher permittivity, small deviations from the light line is observed for the thinnest film ($d = 10$ nm) at high frequency mainly. This mode ($n = 3$) exhibits the higher wave vector but the shorter propagation length with respect to mode $n = 1$, as shown in Fig. 6.3b. The values of Λ increase with the film thickness and monotonically decrease as the frequency increases. In contrast to the behavior of β_R, the saturation of Λ appears for film thicker than 50 nm only. This behavior holds for both SPP modes and indicates that the coupling of SPPs propagating along the film interfaces reduces their propagation length, for films thinner than 50 nm. This destructive coupling is generated by the permittivity difference between the substrate and superstrate ($\varepsilon_3 - \varepsilon_1 = 10.7$), which

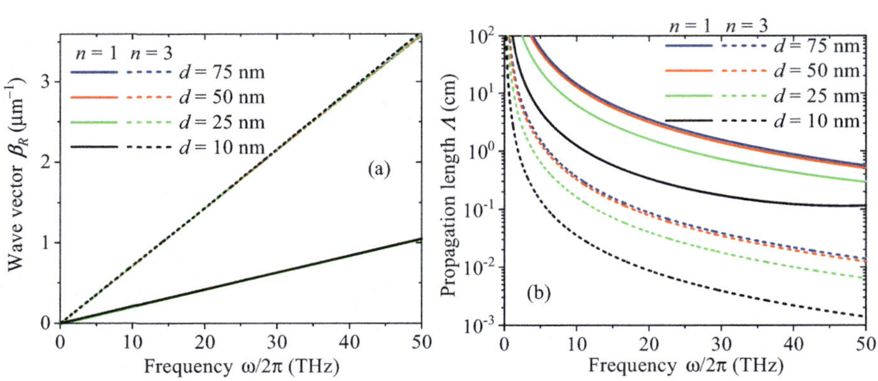

Fig. 6.3 Spectra of the in-plane (**a**) wave vector and (**b**) propagation length of SPPs propagating along the interfaces with air ($n = 1$) and Si ($n = 3$) of a gold nanofilm. Calculations were done for four representative film thicknesses. The continuous and dashed lines in (**a**) are pretty much overlapping with their corresponding black ones

increases the SPP energy loss as the film thickness decreases. By contrast, for a film surrounded up and down by the same material, the SPP energy loss is lower for thinner (less absorbing) films, and therefore, Λ increases when the film thickness reduces, as shown in Chap. 4.

Note that the SPP propagation along the gold nanofilm satisfies the condition $\mathrm{Im}(\beta) \ll \mathrm{Re}(\beta)$, which indicates that the SPP propagation length $\Lambda = [2\mathrm{Im}(\beta)]^{-1} \gg [2\mathrm{Re}(\beta)]^{-1}$ is generally greater than the in-plane wavelength $2\pi/\mathrm{Re}(\beta)$. The relatively long distances ($\Lambda \sim 1$ cm) that the SPPs propagate show their high potential to carry heat along the nanofilm interfaces. The reduction of the propagation length for high frequencies and thinner films is also exhibited by the SPP skin (penetration) depth into the air (δ_1) and Si (δ_3) surrounding the gold nanofilm, as shown in Fig. 6.4a, c, respectively. The absorption of the SPP energy by the film hence reduces both the in-plane and cross-plane propagation distances of SPPs. SPPs with shorter in-plane propagation lengths are thus more confined to the film interfaces. In both surrounding media, the first SPP mode ($n = 1$) penetrates deeper than the second one ($n = 3$), such that they both go deeper into the air than into Si ($\delta_1 > \delta_3$), for a given frequency and film thickness. Further, the penetration

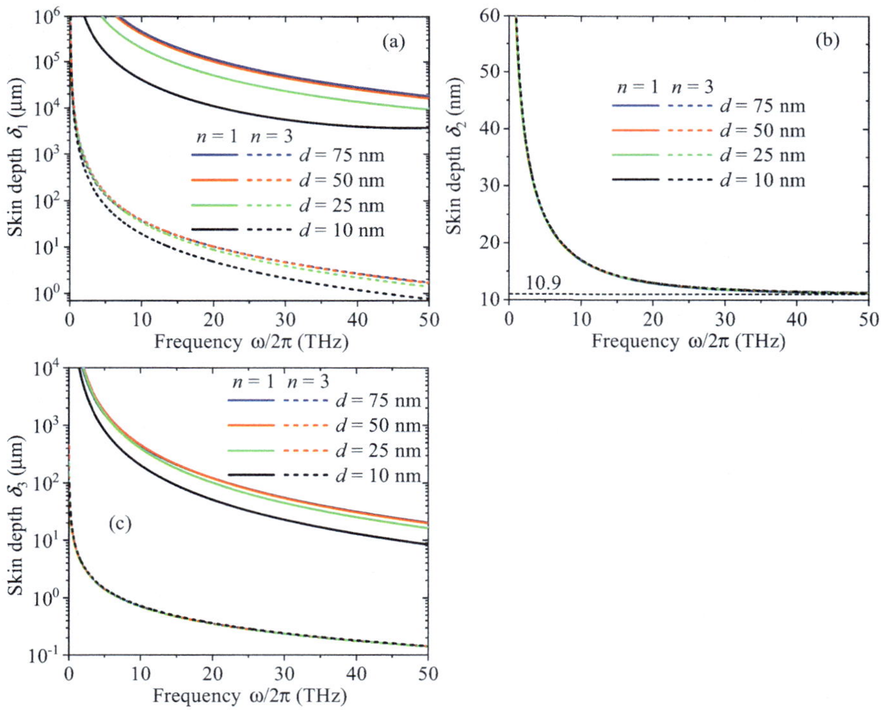

Fig. 6.4 Spectra of the cross-plane propagation lengths (skin depths) of SPPs propagating in (**a**) vacuum (**b**) gold film, and (**c**) its Si substrate, for the modes $n = 1$ and 3. The continuous and dashed lines in (**b**) and the dashed ones in (**c**) are overlapped

depth of the second mode into the Si substrate is independent of the film thickness and exhibits the strongest confinement (smallest skin depth) to the film interface. The skin depth δ_2 of SPPs propagating inside the film also exhibits this thickness independence and takes values much smaller than δ_3, for mode $n = 1$, due to the nanofilm energy absorption, as shown in Fig. 6.4b. These values of δ_2 are the same for both modes and are accurately determined by $\delta_2 = \left[2k_0\text{Re}(\sqrt{-\varepsilon_2})\right]^{-1}$, which confirms the validity of the approximation ($p_2 \approx k_0\sqrt{-\varepsilon_2}$) used to derive and solve the dispersion relation in (6.10). As the frequency increases, δ_2 decreases and tends to $0.5c/\omega_p = 10.9$ nm, as predicted by (6.20). For films thicker than 50 nm, $d > \delta_2$ for the vast majority of frequencies and therefore the SPPs propagating along both film interfaces are decoupled. This fact indicates that the SPP propagation becomes independent of the film thickness for thicknesses $d > 50$ nm. For thinner films, on the other hand, the intrafilm skin depth can be greater or smaller than the film thickness, within a representative frequency interval, and hence the coupling (thickness) effects show up, as shown in Figs. 6.3 and 6.4.

The contributions of the two SPP modes to the spectrum G_ω/a of the SPP thermal conductance per unit width ($G = \int G_\omega d\omega$) of the gold nanofilm are shown in Fig. 6.5a, for four representative film thicknesses. The maxima of G_ω and their frequencies of occurrence increase with the film thickness until reaching saturation for $d > 50$ nm. This saturation is related to the one of the propagation length and arises from the decoupling ($d > \delta_2$) of the plasmon modes propagating along the two film interfaces, as shown in Fig. 6.4b. These peak frequencies maximize both β_R and Λ, and therefore their vicinity provides the major contribution to the SPP thermal conductance. For frequencies higher than a given peak frequency, the sharp reduction of G_ω is driven by the exponential decrease of the Bose–Einstein distribution function, such that its values nearly vanish for $\omega/2\pi > 50$ THz. For a given film thickness, the peak frequency of mode $n = 1$ is higher than that of mode $n = 3$, which has a higher amplitude than the former one, for $d \geq 25$ nm. This fact indicates that both modes have a significant contribution to G, as shown

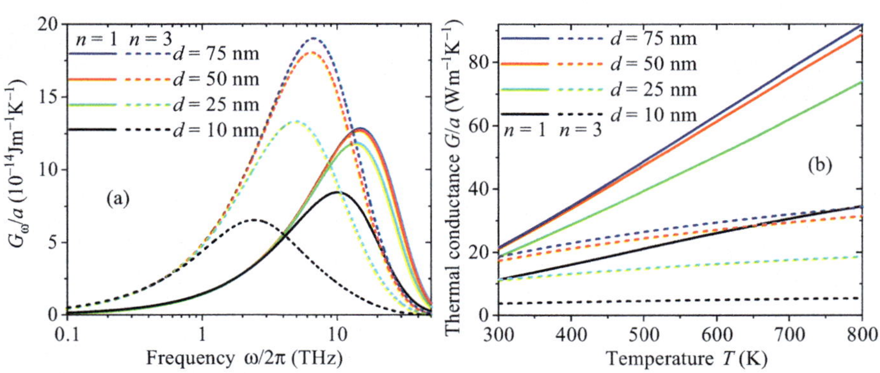

Fig. 6.5 (**a**) Spectrum of the SPP thermal conductance of a 1-cm-long gold nanofilm and its (**b**) integrated counterpart as a function of temperature. Calculations were done for both SPP modes, four film thicknesses, and $T = 300$ K in (**a**)

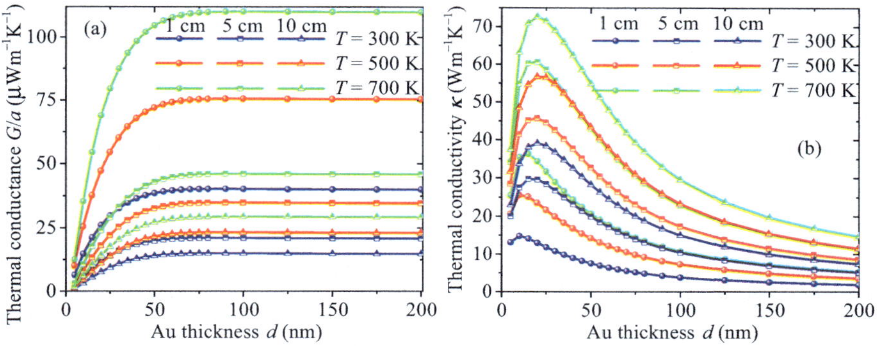

Fig. 6.6 (**a**) SPP thermal conductance per unit width and (**b**) thermal conductivity of a gold nanofilm as functions of its thickness. Calculations were done by summing the contributions of the two SPP modes predicted by (6.9), for film lengths of 1, 5, and 10 cm; and three representative temperatures

in Fig. 6.5b. Note that the contribution of mode $n = 1$ is generally higher than the one of mode $n = 3$, especially at high temperature. The near-linear increase of G with temperature deviates from the quadratic one predicted by (6.23) for the ballistic regime ($l \ll \Lambda$) because the propagation length shown in Fig. 6.3b takes smaller and greater values than the film length $l = 1$ cm considered for plotting Fig. 6.5b. Therefore, in the intermediate ballistic–diffusive regime, the SPP thermal conductance increases with temperature slower than in the pure ballistic one.

The sum of the contributions of both SPP modes to the SPP thermal conductance G/a per unit width of a gold nanofilm is shown in Fig. 6.6a, as a function of its thickness d. Note that G increases with d until reaching a plateau for $d > 50$ nm, due to the uncoupling ($d > \delta_2$) of the SPP modes propagating along the film interfaces. This maximum G value increases with the film length and temperature, as a result of the longer propagation distance in a wider frequency range. The strong enhancement of G with temperature is driven by the relatively board frequency spectrum (from 0 to 50 THz mainly) supporting the flow of the SPP thermal energy, as shown in Fig. 6.5a. Given that the upper bound of this frequency range is much smaller than the gold plasma frequency (50 THz$\ll \omega_p/2\pi = 2196.34$ THz), the frequency spectrum of the SPP thermal conductance is not limited by ω_p, which results in the absence of the saturation of G at high temperature. For thinner films ($d < 50$ nm), on the other hand, G decreases with d as a consequence of the destructive coupling of SPPs propagating along the film interfaces. In terms of the SPP thermal conductivity $k = Gl/ad$, the coupling–uncoupling transition of G appears as a maximum due to its division by the film thickness d. This maximum appears for a film thickness increasing with temperature and film length, as shown in Fig. 6.6b. For a 1-cm-long nanofilm at 300 K, a maximum thermal conductivity of 15 $\mathrm{Wm^{-1}K^{-1}}$ appears at the thickness of 10 nm. This SPP thermal conductivity is about 25% of its electron counterpart [4] and takes higher values for longer and/or hotter films, without increasing its thickness ($d < 25$ nm) significantly. The sizable increase of the SPP thermal conductivity with the relatively long film length, while its electron counterpart keeps independent of it, indicates that the SPPs can be powerful heat carriers able to enhance the in-plane heat transport of metallic films, especially at temperatures higher than room temperature.

6.2 Plasmon Heat Transport Driven by Graphene

Taking into account that the SPPs propagate along interfaces mainly, their contribution to the heat conduction could naturally be enhanced via highly conductive 2D materials, such as graphene. The well-known ability of graphene to support plasmon modes has the potential to enhance the thermal performance of solids coated with graphene, without changing their internal structure. To quantify this enhancement, we consider a suspended film supporting the propagation of SPPs along its two interfaces coated with graphene monolayers, as shown in Fig. 6.7. The film and its surrounding medium have relative permittivities ε_2 and ε_1, respectively. Considering that the film surfaces $z = 0$ and $z = l$ are in contact with thermal baths at temperatures T_h and $T_c(< T_h)$, the free electrons of graphene oscillate and emit an electromagnetic field (SPP) that propagates along the film interfaces mainly. Under this thermal excitation, the heat propagates along the z axis via the combined dynamics of the film phonons (or electrons) and SPPs. Given that SPPs are bosons, their contribution k to the in-plane thermal conductivity of the film can be determined through (5.34). As the predictions of (5.34) are driven by the real and imaginary parts of the in-plane wave vector β of SPPs, we derive their dispersion relation for the long-range modes predicted by the TM polarization.

The tangential components of the TM fields propagating below $(E_z^<, H_y^<)$, inside (E_z, H_y), and above $(E_z^>, H_y^>)$ the film shown in Fig. 6.7 are determined by (4.2) and (4.3), which yield

$$E_z^< = A_< e^{p_1(x+d/2)}, \quad H_y^< = -\frac{i\omega\varepsilon_0\varepsilon_1}{p_1} E_z^<, \tag{6.29}$$

$$E_z = A\cosh(p_2 x) + B\sinh(p_2 x), \quad H_y = -\frac{i\omega\varepsilon_0\varepsilon_2}{p_2^2}\frac{\partial E_z}{\partial x}, \tag{6.30}$$

$$E_z^> = A_> e^{-p_1(x-d/2)}, \quad H_y^> = \frac{i\omega\varepsilon_0\varepsilon_1}{p_1} E_z^>, \tag{6.31}$$

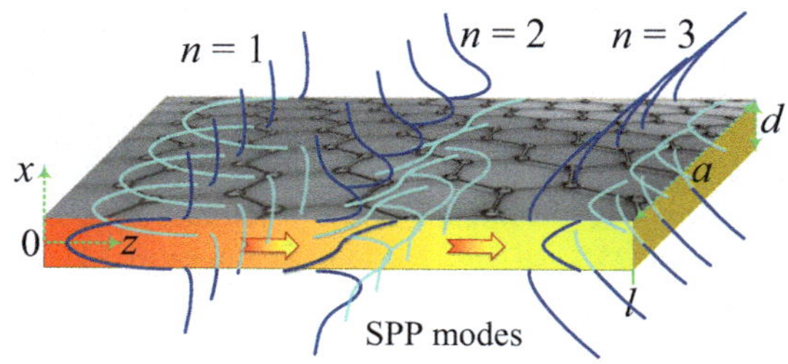

Fig. 6.7 Scheme of a film supporting the propagation of three SPP modes. The curvy lines stand for the SPP magnetic field, whose discontinuity at the film interfaces appears due to their coating with a graphene monolayer

where $p_n^2 = \beta^2 - \varepsilon_n k_0^2$ and the solutions $\exp[p_1(x + d/2)]$ and $\exp[-p_1(x - d/2)]$ were chosen to ensure the exponential decay of the fields as $x \to -\infty$ and $x \to +\infty$, respectively, provided that $\mathrm{Re}(p_1) > 0$. According to the boundary conditions in (3.79) and (3.80), the four parameters $A_<$, $A_>$, A, and B are determined by the continuity of the electric field component (E_z) and the discontinuity of the magnetic field component (H_y) at the boundary surfaces $x = \pm d/2$. By describing the surface current density J_{Sz} of graphene by means of the Ohm's law $J_{Sz} = \sigma E_z$, with σ being the graphene electric conductivity, these conditions takes the form

$$E_z^<(x = -d/2) = E_z(x = -d/2), \quad E_z^>(x = d/2) = E_z(x = d/2), \tag{6.32}$$

$$H_y^<(x = -d/2) - H_y(x = -d/2) = \sigma E_z^<(x = -d/2), \tag{6.33}$$

$$H_y(x = d/2) - H_y^>(x = d/2) = \sigma E_z^>(x = d/2). \tag{6.34}$$

After inserting (6.29)–(6.31) into (6.32)–(6.34), one obtains

$$A_< = A\cosh(\alpha) - B\sinh(\alpha) = \frac{\delta_2}{\delta_1 + i\sigma/\omega\varepsilon_0}[-A\sinh(\alpha) + B\cosh(\alpha)], \tag{6.35}$$

$$A_> = A\cosh(\alpha) + B\sinh(\alpha) = -\frac{\delta_2}{\delta_1 + i\sigma/\omega\varepsilon_0}[A\sinh(\alpha) + B\cosh(\alpha)], \tag{6.36}$$

where $\alpha = p_2 d/2$ and $\delta_n = \varepsilon_n/p_n$. The solution of (6.35) and (6.36) can easily be obtained by summing and subtracting them, as follows:

$$A_< + A_> = 2A\cosh(\alpha) = -\frac{2\delta_2}{\delta_1 + i\sigma/\omega\varepsilon_0} A\sinh(\alpha), \tag{6.37}$$

$$A_< - A_> = -2B\sinh(\alpha) = \frac{2\delta_2}{\delta_1 + i\sigma/\omega\varepsilon_0} B\cosh(\alpha). \tag{6.38}$$

Equations (6.37) and (6.38) have two possible solutions, and therefore they predict the existence of the following two modes:

Symmetric Mode ($B \neq 0$)
According to (6.37) and (6.38), this mode is characterized by $A = 0$, $A_> = -A_< = B\sinh(\alpha)$ and the following dispersion relation and fields

$$\frac{\varepsilon_1}{p_1} + \frac{\varepsilon_2}{p_2}\coth(p_2 d/2) = -\frac{i\sigma}{\omega\varepsilon_0}, \tag{6.39}$$

$$E_z^< = -B\sinh(p_2 d/2)\, e^{p_1(x+d/2)}, \tag{6.40}$$

$$E_z = B\sinh(p_2 x), \tag{6.41}$$

$$E_z^> = B\sinh(p_2 d/2)\, e^{-p_1(x-d/2)}. \tag{6.42}$$

The combination of (6.30) and (6.41) thus establishes that the magnetic field is an even (symmetric) function of position ($H_y(-x) = H_y(x)$), which justifies the mode name.

Antisymmetric Mode ($A \neq 0$)
In this case, (6.37) and (6.38) predict that $B = 0$, $A_> = A_< = B\cosh(\alpha)$, and

$$\frac{\varepsilon_1}{p_1} + \frac{\varepsilon_2}{p_2}\tanh(p_2 d/2) = -\frac{i\sigma}{\omega\varepsilon_0}, \tag{6.43}$$

$$E_z^< = A\cosh(p_2 d/2)\, e^{p_1(x+d/2)}, \tag{6.44}$$

$$E_z = A\cosh(p_2 x), \tag{6.45}$$

$$E_z^> = A\cosh(p_2 d/2)\, e^{-p_1(x-d/2)}. \tag{6.46}$$

According to (6.30) and (6.45), the magnetic field of this mode is an odd (antisymmetric) function of position ($H_y(-x) = -H_y(x)$), which gives the mode name. Note that, in the absence of graphene ($\sigma = 0$), (6.39) and (6.43) reduce to the dispersion relations of the even and odd modes for an uncoated film, respectively. For a very thick film ($|p_2|\, d \gg 2$), both (6.39) and (6.43) become independent of the film thickness and reduce to the dispersion relation $\varepsilon_1/p_1 + \varepsilon_2/p_2 = -i\sigma/(\omega\varepsilon_0)$ of a single interface. For very thin films ($|p_2|\, d \ll 2$), on the other hand, the solutions of (6.39) and (6.43) for $\beta = k_0\sqrt{\varepsilon}$ can conveniently be found in terms of the effective permittivity ε and the normalized thickness $\lambda = k_0 d/2$. The cross-plane wave vectors then read $p_n = k_0\sqrt{\varepsilon - \varepsilon_n}$, for $n = 1$ and 2; and (6.39) and (6.43) take the following dimensionless forms:

$$\frac{\varepsilon_1}{\sqrt{\varepsilon - \varepsilon_1}} + \frac{\varepsilon_2}{\sqrt{\varepsilon - \varepsilon_2}}\coth\left(\lambda\sqrt{\varepsilon - \varepsilon_2}\right) = -iS, \tag{6.47}$$

$$\frac{\varepsilon_1}{\sqrt{\varepsilon - \varepsilon_1}} + \frac{\varepsilon_2}{\sqrt{\varepsilon - \varepsilon_2}}\tanh\left(\lambda\sqrt{\varepsilon - \varepsilon_2}\right) = -iS, \tag{6.48}$$

where $S = \sigma/c\varepsilon_0$ is the normalized and dimensionless electrical conductivity of graphene. Assuming that the film thickness and frequency are small enough to make $\lambda \ll 1$, the Taylor series expansion of $\coth\left(\lambda\sqrt{\varepsilon - \varepsilon_2}\right) \approx \left(\lambda\sqrt{\varepsilon - \varepsilon_2}\right)^{-1} + \lambda\sqrt{\varepsilon - \varepsilon_2}/3$ and $\tanh\left(\lambda\sqrt{\varepsilon - \varepsilon_2}\right) \approx \lambda\sqrt{\varepsilon - \varepsilon_2}$, for an approximation up to λ^2, leads to

$$\frac{\varepsilon_1}{\sqrt{\varepsilon - \varepsilon_1}} + \frac{\varepsilon_2}{\lambda(\varepsilon - \varepsilon_2)}\left[1 + \frac{\lambda^2}{3}(\varepsilon - \varepsilon_2)\right] = -iS \tag{6.49}$$

$$\frac{\varepsilon_1}{\sqrt{\varepsilon - \varepsilon_1}} + \varepsilon_2\lambda = -iS. \tag{6.50}$$

The dependence on λ of (6.49) and (6.50) indicates that ε has the following general solution:

$$\varepsilon = \frac{a_{-2}}{\lambda} + \frac{a_{-1}}{\sqrt{\lambda}} + a_0 + a_1\sqrt{\lambda} + a_2\lambda + a_3\sqrt{\lambda}^3 + a_4\lambda^2 + \ldots, \tag{6.51}$$

where the coefficients a_m are obtained by replacing (6.51) into (6.49) and (6.50). The solutions $n = 1$ and $n = 2$ thus obtained for the dispersion relations of the symmetric and antisymmetric modes supported by graphene ($S =\neq 0$) are, respectively, given by

$$\varepsilon = \frac{i\varepsilon_2}{S\lambda} + \frac{i\varepsilon_1}{S}\sqrt{\frac{i\varepsilon_2}{S\lambda}} + \varepsilon_2 - \frac{1}{S^2}\left(\frac{\varepsilon_2^2}{3} + \frac{\varepsilon_1^2}{2}\right), \tag{6.52}$$

$$\varepsilon = \varepsilon_1 - \left(\frac{\varepsilon_1}{S}\right)^2\left[1 + 2i\frac{\varepsilon_2\lambda}{S} - 3\left(\frac{\varepsilon_2\lambda}{S}\right)^2\right]. \tag{6.53}$$

Given that $\lambda \ll 1$, the leading term in (6.52) indicates that the mode $n = 1$ features a large wave vector $\mathrm{Re}(\beta) \to d^{-1/2}$ and a small propagation length $\Lambda \to d^{1/2}$. These short-wavelength and short-range SPPs appear due to the increasing energy absorption of the coating graphene, as the film thickness scales down. For this branch $n = 1$, the SPP propagation along the in-plane and cross-plane directions is driven by the same wave vectors ($\beta \approx p_1 \approx p_2$). For mode $n = 2$, on the other hand, $\beta \to k_0\sqrt{\varepsilon_1 - (\varepsilon_1/S)^2}$ becomes independent of the thickness and permittivity of the film, for thin enough films. Therefore, the coupling of the SPP mode $n = 2$ propagating along the two film interfaces saturates, as the film thickness sizes down. Note that both modes $n = 1$ and $n = 2$ exist due to the presence of graphene ($S \neq 0$), and they are respectively driven by the media permittivities $\varepsilon \to \varepsilon_2$ and $\varepsilon \to \varepsilon_1$, for a very high conductivity $S \to \infty$. Both modes thus exist even if the film and its surrounding medium are made of the same material ($\varepsilon_1 = \varepsilon_2$). However, as a result of the permittivity difference $\varepsilon_1 - \varepsilon_2 \neq 0$, the combination of (6.49) and (6.51) predicts the propagation of a third mode ($n = 3$) determined by

$$\varepsilon = \varepsilon_2 + \left(1 - \frac{\varepsilon_2}{\varepsilon_1}\right)^2 (\varepsilon_2\lambda)^2\left[1 + 2i\left(1 - \frac{\varepsilon_2}{\varepsilon_1}\right)S\lambda\right]. \tag{6.54}$$

The leading dependence of this third mode on the graphene conductivity is thus weighted by the normalized thickness λ to the third power. This weak dependence of ε on S indicates that the mode $n = 3$ is not strongly affected by the graphene energy absorption, and therefore it can propagate long distances, in the absence of energy absorption by the film and its surrounding medium ($\mathrm{Im}(\varepsilon_2) = \mathrm{Im}(\varepsilon_1) = 0$). As the SPP thermal conductivity increases with the SPP propagation length, this third mode is expected to significantly enhance the heat transport along the film interfaces. Therefore, while the propagation of the first two modes is driven by either the permittivity of the film or its surrounding medium, the third mode exists ($p_2 = k_0\sqrt{\varepsilon - \varepsilon_2} \neq 0$) due to the permittivity difference of these media.

We now determine numerically the propagation and thermal properties of SPPs propagating along an Si film ($\varepsilon_2 = 11.7$) surrounded by air ($\varepsilon_1 = 1$). These two media do not absorb energy in a large frequency range, and therefore they favor the propagation of SPPs over long distances. The graphene monolayers on the top and bottom surfaces of the film are good SPP conductors that absorb energy and hence support the SPP heat transport. For temperatures comparable to or higher than 300 K, the electric conductivity σ of a graphene monolayer can be described by the following Drude-like model:

$$\sigma = \frac{\Omega}{\gamma - i\omega}, \tag{6.55}$$

where γ is the electron scattering rate and $\Omega = (e/\hbar)^2 E_F/\pi$, with e and E_F being the electron charge and Fermi energy, respectively. σ is thus determined by two parameters, whose experimental values are $\gamma/2\pi = 1.59$ THz and $E_F = 1$ eV [5, 6]. Equation (6.55) represents the intraband contribution to σ, as the interband one becomes negligible for $\hbar\omega < 2E_F$ ($\omega/2\pi < 483$ THz). This frequency interval covers the vast majority of relevant frequencies for the propagation and heat conduction of plasmons, as shown below for temperatures between 300 K and 700 K.

The in-plane wave vector β_R and propagation length Λ of SPPs propagating along a graphene-coated Si film are shown in Fig. 6.8a, b, respectively. Mode $n = 1$ exhibits the largest β_R (shortest wavelength) and smallest Λ, which takes values below 30 μm. The wave vector of mode $n = 1$ increases with frequency and takes higher values for thinner films, while its propagation length exhibits the opposite behavior. This short-wavelength mode ($n = 1$) is therefore expected to contribute to the SPP heat transport due to its large wave vector, which is nearly two orders of magnitude larger than the corresponding one of mode $n = 3$. For mode $n = 3$, β_R linearly increases with frequency and is very close to the wave vector of light

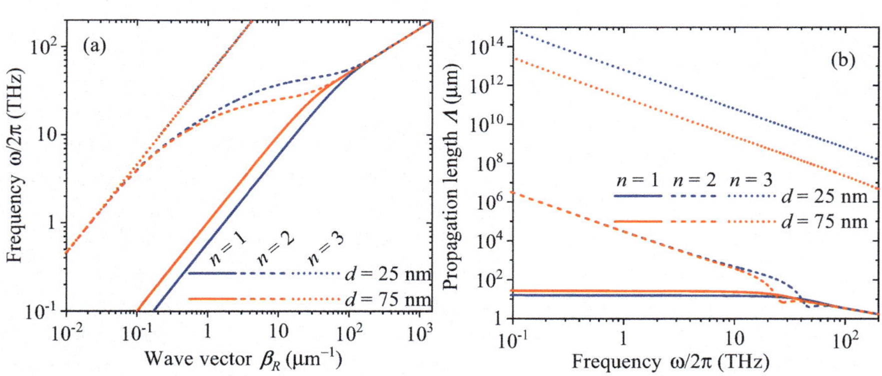

Fig. 6.8 Spectra of the in-plane (**a**) wave vector and (**b**) propagation length of the 3 SPP modes propagating along a graphene-coated Si film with a thickness $d = 25$ and 75 nm

Fig. 6.9 Group velocity $V_g = \partial\omega/\partial\beta_R$ of the SPP modes, as a function of frequency. Calculations were done for the in-plane wave vector shown in Fig. 6.8a

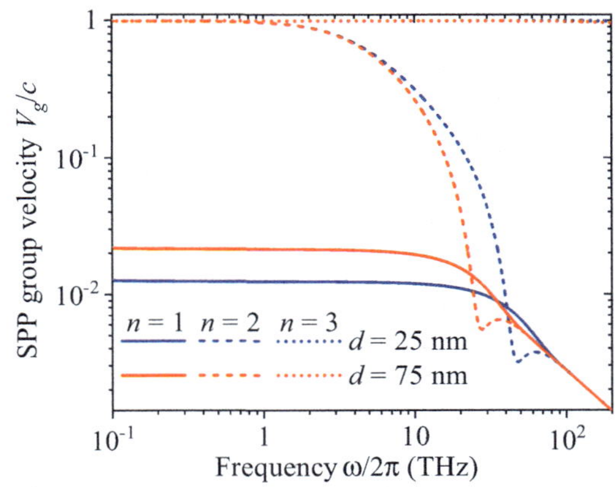

in vacuum ($\beta_R \approx k_0$). This photon-like nature of SPPs is accompanied by very long propagation lengths resulting from the weak energy absorption of graphene. The major contribution of mode $n = 3$ to the SPP heat transport is thus expected to arise from its long propagation length. Mode $n = 2$, on the other hand, has intermediate wave vectors and propagation lengths, whose values are between the corresponding ones of the other two modes. Unlike mode $n = 1$, the propagation length of mode $n = 2$ is nearly independent of the film thickness, for frequencies lower than 10 THz. According to Fig. 6.9, the three modes propagate with a group velocity near to the speed of light in vacuum, and hence they can drive a plasmonic heat conduction much faster than those of phonons and electrons.

The propagation length of the three SPP modes has a similar behavior than the SPP skin depth $\delta_2 = [2\text{Re}(p_2)]^{-1}$ inside the Si film, as shown in Fig. 6.10a. Therefore, the energy absorption by the graphene-coated film reduces both the in-plane (Λ) and cross-plane (δ_1) propagation distances of SPPs. For a given frequency, the intrafilm skin depth satisfies the relation $\delta_2(n = 1) < \delta_2(n = 2) < \delta_2(n = 3)$, and its values are generally longer than the film thickness ($\delta_2 > d$). The SPPs propagating along the bottom and top film surfaces are thus coupled and exist in a broad band of frequencies. This existence is further confirmed by the spatial distribution of the Poynting vector shown in Fig. 6.10b. The energy density of modes $n = 1$ and 2 is confined along the film interfaces mainly, while the intrafilm energy of mode $n = 3$ is much smaller than the extra-film one, due to the high sensitivity of the wave vector $p_1 = k_0\sqrt{\varepsilon - \varepsilon_1}$ of this third mode to the permittivity difference $\varepsilon_2 - \varepsilon_1 = 10.7$ of the nonabsorbing Si and air, as established by (6.54).

The contributions of the three SPP modes to the spectrum k_ω of the SPP thermal conductivity $k = \int k_\omega d\omega$ of the graphene-coated Si film are shown in Fig. 6.11a. The maxima of k_ω and their frequencies of occurrence increase as the film thickness decreases, such that $k_\omega(n = 1) > k_\omega(n = 3) > k_\omega(n = 2)$. The short-wavelength mode ($n = 1$) thus carries more thermal energy than the long-range one ($n = 3$), as a result of the limited length ($l = 1$ mm) of the considered film. Mode $n = 2$ exhibits the lowest contribution because the peak of k_ω results from the maximization of

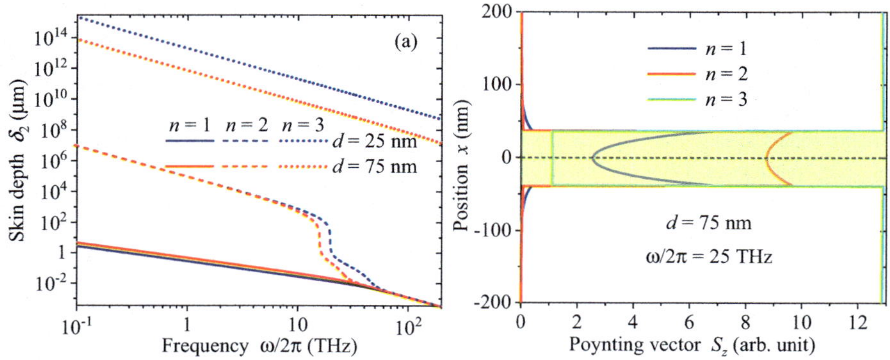

Fig. 6.10 (**a**) Frequency spectrum of the cross-plane propagation length of the three SPP modes and (**b**) spatial distribution of the in-plane component of their Poynting vector

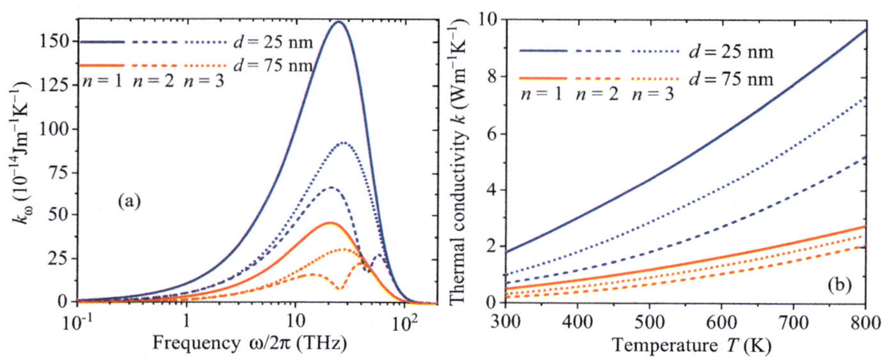

Fig. 6.11 (**a**) Frequency spectrum of the SPP thermal conductivity of a graphene-coated Si film and its (**b**) integrated counterpart as a function of temperature. Calculations were done for the three SPP modes, two film thicknesses, $l = 1$ mm, and $T = 300$ K in (**a**)

both β_R and Λ, which take intermediate values only, as shown in Fig. 6.8. For frequencies higher than the peak ones, the sharp reduction of k_ω is driven by the exponential decrease of the Bose–Einstein distribution function, such that its values nearly vanish for $\omega/2\pi > 100$ THz. The three SPP modes thus exhibit broad frequency spectra and their contribution to the SPP thermal conductivity fulfill the relation $k(n = 1) > k(n = 3) > k(n = 2)$, as shown in Fig. 6.11b. The near-linear increase of k with temperature deviates from the quadratic one for the ballistic regime ($l \ll \Lambda$) because the propagation length of modes $n = 1$ and $n = 2$ takes smaller and greater values than the film length $l = 1$ mm.

The sum of the contributions of the three SPP modes to the SPP thermal conductivity k of the graphene-coated Si film is shown in Fig. 6.12a, as a function of its thickness d. Note that k monotonically increases as d decreases following nearly the power law $k \propto 1/d$. The plasmon thermal conductivity therefore exhibits

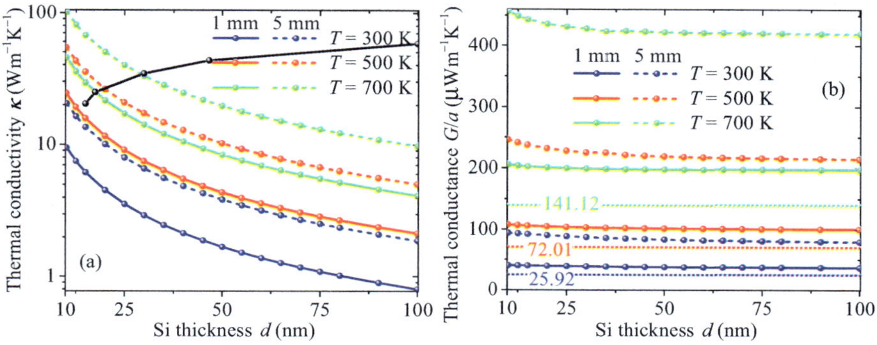

Fig. 6.12 (**a**) SPP thermal conductivity and (**b**) thermal conductance per unit width of a graphene-coated Si film, as functions of its thickness. Calculations were done by summing the contributions of the three SPP modes, for film lengths of 1 and 5 mm; and three representative temperatures. The black points in (**a**) stand for the phonon thermal conductivity of a suspended Si film at T = 300 K [7], while the dotted lines in (**b**) represent the quantum of thermal conductance $G_0/a = 12z(3)k_B^3T^2/ch^2$ of a polar nanofilm

an opposite thickness dependence to its phonon counterpart (black line) reported for a bare Si membrane at 300 K [7]. $k = 13.6$ Wm^{-1}K^{-1} for an Si film at 300 K and a length (thickness) of 5 mm (15 nm). This SPP thermal conductivity represents 67% of its phonon counterpart for a bare Si film. As higher k values are obtained for longer and/or hotter films, the sizable plasmon heat conduction driven by graphene has the potential to counterbalance the reduction of the phonon counterpart, as the film length or temperature increases. In terms of the SPP thermal conductance $G = kad/l$, the SPP contribution is (13.6 Wm^{-1}K^{-1})$\times$ (15 nm)$\times a/l = 204 \times (a/l)$ nWK^{-1}, for a graphene–Si–graphene sandwich at 300 K and a length (thickness) of 5 mm (15 nm). On the other hand, considering that the thermal conductivity of a supported graphene monolayer (thickness $\approx$ 0.4 nm) lies between 300 and 600 Wm^{-1} K^{-1} [8–10], the estimated upper bound for the phonon+electron thermal conductance of the same structure is [(600 Wm^{-1} K^{-1})$\times$(2 $\times$ 0.4 nm) + (20.3 Wm^{-1} K^{-1})$\times$(15 nm)]$\times a/l = 784.5 \times (a/l)$ nWK^{-1}. The SPP contribution to the heat current is hence (204/784.5) $\times$ 100% = 26% of its electron+phonon counterpart.

Taking into account that the SPP heat conduction involves the contribution of the long-range mode $n = 3$, which exhibits a ballistic propagation ($l \ll \Lambda$), for any film length of practical interest ($l < 1$ m), its contribution to the SPP thermal conductivity increases linearly with the film length ($k \propto l$). The SPP thermal conductance of this ballistic mode is therefore independent of the film length. Consequently, the SPP contribution to the heat current is expected to become comparable to or even higher than that of phonons and electron, as the sandwich structure length increases. The behavior $k \propto 1/d$ yields a significant SPP thermal conductance nearly independent of the film thickness, as shown in Fig. 6.12b. This saturation of G arises from the opposite dependence on d of the SPP wave vector and propagation length, and therefore their impact on G nearly cancels out for each mode and film thickness ($d < 100\,\text{nm}$). For a given temperature, G increases with the film length, and its values are higher than the maximum one G_0 (dotted lines) supported by an uncoated polar nanofilm. This $G > G_0$ results from the combined contribution of three plasmon modes, while G_0 is determined by the single mode supported by a polar nanofilm. The graphene coating therefore transforms a Si nanofilm into a plasmonic conductor able to enhance the in-plane heat conduction better than SPhPs in a polar nanofilm, for a given thickness and temperature.

References

1. Ordonez-Miranda, J., Tranchant, L., Tokunaga, T., Kim, B., Palpant, B., Chalopin, Y., Antoni, T., Volz, S.: Anomalous thermal conductivity by surface phonon-polaritons of polar nanofilms due to their asymmetric surrounding media. J. Appl. Phys. **113**, 084311 (2013)
2. Agranovich, V.M.: Surface Polaritons. Elsevier, Amsterdam (2012)
3. Palik, E.D.: Handbook of Optical Constants of Solids. Academic Press, Orlando (1985)
4. Chen, G., Hui, P.: Thermal conductivities of evaporated gold films on silicon and glass. Appl. Phys. Lett. **74**, 2942 (1999)
5. Koppens, F.H.L., Chang, D.E., Garcia de Abajo, F.J.: Graphene plasmonics: a platform for strong light–matter interactions. Nano Lett. **11**, 3370–3377 (2011)
6. Guizal, B., Antezza, M.: Light-induced optomechanical forces in graphene waveguides. Phys. Rev. B **93**, 115427 (2016)
7. Cuffe, J., Eliason, J.K., Maznev, A.A., Collins, K.C., Johnson, J.A., Shchepetov, A., Prunnila, M., Ahopelto, J., Sotomayor Torres, C.M., Chen, G., Nelson, K.A.: Reconstructing phonon mean-free-path contributions to thermal conductivity using nanoscale membranes. Phys. Rev. B **91**, 245423 (2015)
8. Cai, W., Moore, A.L., Zhu, Y., Li, X., Chen, S., Shi, L., Ruoff, R.S.: Thermal transport in suspended and supported monolayer graphene grown by chemical vapor deposition. Nano Lett. **10**, 1645–1651 (2010)
9. Sadeghi, M.M., Jo, I., Shi, L.: Phonon-interface scattering in multilayer graphene on an amorphous support. PNAS **110**, 16321–16326 (2013)
10. Seol, J.H., Moore, A.L., Shi, L., Jo, I., Yao, Z.: Thermal conductivity measurement of graphene exfoliated on silicon dioxide. J. Heat Transfer **133**, 022403 (2010)

Index

S. Volz, J. Ordonez-Miranda, *Heat Transport Driven by Surface Electromagnetic Waves*, Mechanical Engineering Series, https://doi.org/10.1007/978-3-031-66024-5

SPRINGER NATURE

GPSR Compliance

The European Union's (EU) General Product Safety Regulation (GPSR) is a set of rules that requires consumer products to be safe and our obligations to ensure this.

If you have any concerns about our products, you can contact us on ProductSafety@springernature.com

In case Publisher is established outside the EU, the EU authorized representative is:

Springer Nature Customer Service Center GmbH
Europaplatz 3
69115 Heidelberg, Germany

The manufacturer's authorised representative in the EU is Springer Nature Customer Service Centre GmbH, Europaplatz 3, 69115 Heidelberg, Germany. If you have any concerns regarding our products, please contact ProductSafety@springernature.com

Printed and bound by CPI Group (UK) Ltd, Croydon, CR0 4YY
01/10/2026
02225883-0008